HOME WORK

Handbuilt Shelter

Lloyd Kahn

Shelter Publications · Bolinas, California

Distributed in the United States and Canada by Publishers
Group West.

Library of Congress Cataloging-in-Publication Data

Kahn, Lloyd, 1935–
  Home work : handbuilt housing / by Lloyd Kahn.
    p.   cm.
Sequel to Shelter. 1973.
Includes bibliographical references and index.
   ISBN-13: 978-0-936070-33-9 (trade paperback)
   ISBN-10: 0-936070-33-1 (trade paperback)
  1. Dwellings — Pictorial works.   2. House construction.
  3. Dwellings — Design and construction — Amateurs'
  manuals.   4. Architects and builders — Interviews.
  5. Dwellings — Pictorial works.   6. Architecture, Domestic.
  7. Vernacular architecture.   I. Title.
TH4815 .K34 2004
690'.837 — dc22

                                        2003018478

9 8 7 6 5 — 09 08 07

(Lowest digits indicate number and year of latest printing.)

Printed in China

Additional copies of this book may be purchased at your
favorite bookstore or by sending $26.95 plus
$5.50 shipping and handling charges to:

Shelter Publications, Inc.
P. O. Box 279
Bolinas, California 94924
415-868-0280
Email: homework@shelterpub.com
Orders, toll-free: 1-800-307-0131

Visit Our Website
SHELTER ONLINE
www.shelterpub.com

Shelter is more than a roof overhead.

CONTENTS

BUILDERS 1

HOMES 31

NATURAL MATERIALS 73

PHOTOGRAPHERS 97

FANTASY 121

INTRODUCTION

I N THE SUMMER OF 1973, Bob Easton and I produced the book *Shelter*. It was an oversized compendium of buildings and builders around the world and throughout history, containing over 1000 photographs and 250 drawings. It was about doing things for yourself, and doing so efficiently, ecologically, and artistically. It featured people who had created handbuilt homes, and included buildings not seen anywhere else. The book had a feeling of home, hearth, and ingenuity that seemed to capture the spirit of the times. It was picked up by the countercultural underground, became a hit, and is still in print, some 250,000 copies later.

It's been thirty years since *Shelter*, and although our publishing company has gone on to other projects and subjects since, I've stayed interested in building — shooting photos and interviewing builders wherever I've travelled, and collecting books and data on building. *Home Work* is the

result, a summary of what I've found over three decades, and is a sequel to *Shelter*. It's also a sequel in another sense. By a neat twist of karma, it includes a number of people who were inspired by *Shelter* to build homes, and whose lives were changed accordingly. Over the years, a surprising number of people have told us that it inspired them to build something; it gave them the courage to get started.

It may be obvious that a thread of the '60s runs through *Home Work*. Many of these people were motivated by what happened in the '60s. (I certainly was!) In the spirit of the times, they went out and built homes, and they were success-ful — here was a part of the '60s that worked. I started building in the '60s because I needed a place to live and could never find a charming old house to buy. I guess it was my fate; if I wanted a good-feeling home I'd have to create it myself. Over the years, I built four homes, always learning on the job, I found the process of building, and the way things were put together,

fascinating, and I've tried to keep this layman's perspective in gathering information for other owner-builders.

Concurrently with learning to build, I started shooting photos of buildings. I took along cameras and a notebook wherever I travelled, and documented small buildings. Invariably the places that appealed to me most turned out to be owner-built. What was I looking for, what caught my eye? Handmade buildings that did one or more of the following:

- showed good craftsmanship
- were practical, simple, economical, useful
- used resources efficiently
- were tuned into the landscape
- were aesthetically pleasing, radiated good vibes
- showed integrity in design and execution
- (and/or) were wildly creative

Dry rock wall (no mortar),
Wales, 1987

Home Work is not comprehensive in geography — it's heavy on the West Coast, where we live. Nor does it cover all builders, building techniques, or materials. It's country, not city. We haven't tried to cover everything and everyone. It's rather what I've run across over the years, a diverse bunch of buildings, all assembled with human hands.

It's funny — we live in a world powerfully transformed by a number of factors, primarily the digital revolution, yet houses must still be created by hand — your computer's not going to do it for you. We hope *Home Work* will motivate you, will give you the confidence that you can build something if you work at it. A tip: If you're not sure what to do, start!

> *"You never know what's shakin'*
> *until you give it a shake."*
>
> *—Johnny Adams,*
> *blues singer*

What if you can't build a home? Even if that's not in the cards, you can use the ideas (and spirit) here to remodel (or decorate) an apartment, to build a studio, barn, treehouse, workshop, window box, sauna, furniture — to create something with your own hands, with your own body.

There was no master plan in assembling this book. We had a ton of accumulated material — photos, interviews, writing — but no idea what the final result would be. So we just started. We put it together a page at a time, a day at a time. As we went along, the book took on a life of its own. A bunch of this material came in while we were in production, and the book continually changed form. After about a year, *Home Work* seemed to have shaped itself — an organic process of sorts.

Now that it's gone off to the printers, and as I'm writing this, I realize that, along with whatever else *Home Work* is, there is within it a family of builders, a bunch of people around the world with common interests. They're alike in many ways, and they're tuned into many of the same things. Getting them all together in this book allows me to share my discoveries, to show you their work (and to take care of what's become a compulsion to communicate). Hey, Look at what these guys have done!

So, dear reader, come join us on another Shelter journey, an odyssey (in retrospect) of the last thirty years, in this scrapbook of builders, dreamers, and doers — a celebration of the human spirit.

Shelter is more than a roof overhead.

BUILDERS

Master roofer
Stan Thomas

Louie's shop, with poured concrete walls

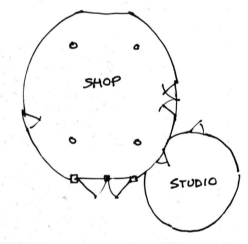

Interior of shop (looking through doors shown above). Structure was inspired by the painting of a Mandan earth lodge in Shelter (at right). This is a classic framing structure, with interior posts and beams providing mid-way support for rafters that span from lower exterior walls. Skylight has two layers of fiberglass: corrugated on outside, flat fiberglass on inside so there's an air space and good insulation.

Crystal at top of mast catches morning sun.

Mandan earth lodge, 1833, as painted by Karl Bodmer

LOUIE FRAZIER

IN THE MID-'80s I went up to the northern California coast to shoot pictures of a house my ex-Bolinas neighbor Jack Williams had built *(see p. 32)*. Jack was a surfer/fisherman/gardener who had had the foresight to get a 39-acre piece of land in Mendocino County in the early '80s. He had built a house and homestead on forested land.

After I shot photographs in the morning, Jack said he had a neighbor who wanted to meet me, who had used our book *Shelter* in building his place, so we drove through hills, then down a winding hillside road into a river-bottom valley. At the end of the road was one of the prettiest little buildings I'd ever seen. Everything about it was right, the curves, the white plastered walls with shingled roof, the copper and crystal mast. We walked up to the open doors of the shop, and a 60-year-old guy, with a handmade hat and a twinkle in his eye came out, a tattered copy of *Shelter* in his hand. "Look," he said beckoning me to squat down with him in the doorway to his shop. He opened the book to the painting of a Mandan earth lodge, and had me look up at the framing of his shop . . . identical!

The quality of Louie's construction was astounding. Everything was beautifully designed, and immaculately carried out. It was all tuned in, thought out, crafted finely. This wasn't *Fine Homebuilding*. This wasn't fussy craftsmanship for millionaires. It was a rare combination of owner-builder-designer-master craftsman, all to a human and livable scale. There was no excess, no fat. This guy made everything: house and shop, chairs and stools, garden cart, cabinets, wood-fired water heaters, hydroelectric system, photovoltaic electricity; he was not only a master carpenter but an arc welder and could figure out how to construct just about anything. He was in the midst of building a beautiful wind-powered fishing boat with his buddy Pete.

Well, that was his shop. And his house? On the other side of the river, and also inspired by a drawing in *Shelter*, was a Japanese-style pole house. To get to it in the winter, you had to ride a bosun's chair across the river on a 500-foot cable *(see p. 8)*.

It was seeing Louie's shop that inspired this book. If *Shelter* could inspire buildings like these, we had better do another book! On these eight pages are some of Louie's creations.

Interior of small circular studio (at right in picture above) attached to shop. Compression ring at peak is '39 Chevy truck wheel rim, rafters nest in the groove of the rim. Sheathing is two layers of ¼" plywood (½" couldn't make the curves).

Top: Louie holding a glass of his homemade Cabernet
Middle: Desk in shop
Bottom: Bedroom in shop before it was converted to studio

more...

Framing shop roof. No floor yet. Walls are covered with foam insulation on outside. Dark vertical and horizontal lines indicate structural grid of building, with four ½" bars of steel. (Insulation is flush with exterior of grid.) See drawings below for details.

"Clothespins" in place, holding form to previously poured course. Foam insulation was then placed inside the form, on the outside of the building before pouring. Each course was 12" high, 8" wide, and 8' long, with two bars of ½" steel. Concrete mixed in gas mixer.

Iron age hut shown in Shelter inspired the walls. "The only time I went up was over a door or window."

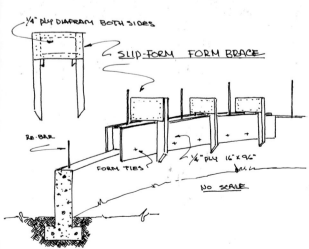

"Large clothespin," used to hold form in place

Bond beam being poured. Note how plywood "large clothespin" slips over previously poured layers.

Roof covered with redwood shakes, Louie and Rufus framing the skylight

Louie's daughter Carrie putting spacers in foundation forms

There's a continuous bond beam with four bars of ½" steel on the top of the wall. Rafters were toenailed to treated 2-by-12's embedded in the bond beam. Concrete was then poured between the rafters.

Shop shown from other side. Studio windows at left

2'x4" RAFTERS
2x6" SLEEPER
BOND BEAM 4 BARS
VERTICAL REBAR 4'O.C.
GALV. FENCING MAT.
1½" TECHFOAM INSL.
½" AIR SPACE
¾" STUCCO TIED TO WALL w/WIRE

TYPICAL SECTION - WALL

more...

100-year-old classic 36" boatbuilder's bandsaw with two wooden wheels

Garden cart with motorcycle wheels, tires, and bearings rides smoothly.

Interior of shop, showing work benches. Stove made from 50-gallon oil drum with cast-iron barrel stove kit.

SHOP TALK

Any builder would love Louie's shop. It's practical, but also bright, cheerful, and aesthetic. Lots of workbench space, an arc-welding setup, elegant old wooden-wheeled band saw, small kitchen, round bedroom attached, CD player, coffee, tequila...

Stool legs made of a quartered fir sapling, stool legs arranged in the same orientation as when sapling was split (see drawing). Seat is piece of foam rubber, covered by piece of old Oriental carpet. Steel strap holds legs together at bottom.

This day Louie looked up at Pete at the other end and said, "Are we rolling, Bob?"

"The easy way is hard enough."
–Sign in
Louie's shop

"My friend Pete and I lost our minds one day and decided to build a boat." They got plans for $2 from the Smithsonian Institute for a 'Crotch Island Pinky,' a sturdy little sailing fishing boat from the East Coast, and embarked on a five-year (unexpectedly long and expensive) odyssey. Framing is white oak; planking (lapstrake) is old-growth clear Douglas fir attached with copper rivets. Deck is teak over marine plywood. Right now Louie is building a cabin and plans to sell the boat. It's designed to use stones for ballast, which can be jettisoned when the boat is loaded with fish. It can get into waters as shallow as two feet. (In case you wondered, Louie and Pete are still good friends.)

 www.royfox.com

The Roy Fox under sail in Sausalito, California

After losing two saunas to high river water, Louie built this one on a one-ton Toyota truck frame. A pickup truck plus a few people haul it back from the river in the winter, with Donna steering the front wheels from inside the sauna. Woodstove built from 50-gallon drum gets fed from outside (on other side). To cool off, you dive into cool, green water.

When I visit Louie, I usually arrive at night and sleep out by the river. Then he shows up at sunrise and we take a sauna. The last time we saw a family of river otters swim through the pool and scuttle over the shallows — they didn't see us inside.

A model of Roy Fox with new cabin, under construction

more...

You could get to Louie's house across a little bridge in the summer. But in the winter when the river was high, you got there on a cable over the river. To demonstrate, we climbed up a steep flight of stairs to a platform about 25 feet off the ground. Louie hooked up a chair, had me sit in it and said to let go. After a first reaction of "no way!," I finally (shakingly) got rolling. It was fabulous. It felt safe and I zinged 500 feet across the river and came into a landing platform. (To get back you unhook the chair, climb up to another tower platform and go back on another cable.)

Pole house on 24 8"-by-8" redwood posts set in ground. Floor of house and decks are 2-by-4's on edge. Walls are redwood 2-by-10's that slide into slots cut by chainsaw in the 8"-by-8" posts. Rafters are curved and there's a wide overhang.

Wood is stacked on deck because the house is in a flood plain.

Louie sitting at base of tower for cable crossing of river (the only way to get across in winter)

Bosun's chair hung from snatch block, with wooden brake shoe. Snatch block locks in place on cable with shackle pin.

View looking down from platform before you take off. Ulp!

Louie coming across river

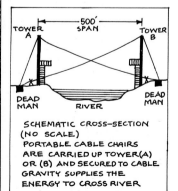

SCHEMATIC CROSS-SECTION (NO SCALE) PORTABLE CABLE CHAIRS ARE CARRIED UP TOWER(A) OR (B) AND SECURED TO CABLE. GRAVITY SUPPLIES THE ENERGY TO CROSS RIVER

Louie's house is a success story of the '60s, owner-designed and built. It's an off-the-grid house, but with amenities. Hot water comes from a coil in the woodstove in winter, from solar heat in summer. In the '80s Louie installed a solar-powered electric system and in the '90s, hydroelectric power (as well as a DirectTV satellite dish).

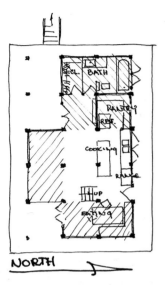

All poles spaced 7' on center each way. Shaded area: sleeping lofts with catwalk connecting.

Sleeping loft. Rafters are 2-by-12's cut to curved shape with chainsaw.

Kitchen is on left, with ladder to sleeping lofts at right.

Bathtub was an abandoned horse trough in a neighbor's pasture.

Black and white photos: Janet Holden Ramos

BALLY HIGH
Ian MacLeod and His
House of Stone in South Africa

I would like to think that Frank Lloyd Wright would have approved. I tried to make my house sympathetic to its surroundings — blending in with the hillside, both in color and shape.

IN 1987 I GOT a blue aerogram from South Africa. It was from Ian MacLeod, a South African of Scottish descent who had singlehandedly built a stone house on a rocky hillside near Warm Baths, S. A. Ian said he'd been inspired by *Shelter* (as well as Frank Lloyd Wright and Simon Rodia) and did I want to see pictures of his house?

In the following years, Ian sent us photos and the story of his building project, wanting to share his adventures with other like-minded people. It was a beautiful house, of local materials, and designed with respect for the earth. It was obvious from his letters that Ian was full of energy, had a sense of humor, and a zest for life (as well as a strong back!). The house was on land owned by a naturist resort, so Ian worked in the nude; the house was so much part of the landscape that inquisitive baboons from across the valley frequently came and jumped up and down on his roof; visitors dropped by to see his creation, and magazines wrote stories about Ian and "The Clan of the Caveman."

Ian and I have been corresponding for almost 15 years now, via letters and the occasional phone call. We share an interest in the art of putting a roof over one's own head. Ian, like Louie Frazier *(see pp. 2–9),* is one of the great inspirations for this book. His spirit, strength, and creativity shine through in his work. Through the years he's kept after me to do *Home Work,* through all the delays and postponements. (I don't think he'll believe this book ever got off the ground until he holds it in his hands.) So here's *Home Work,* Ian; we got it done after all. And here's Ian's great adventure, folks, told in his own words:

What a pleasure to labor with the warmth of the sun on one's back — to feel the cool breeze coping with the good, honest sweat of physical effort, with no sweaty clothing to worry about — and then, at the end of the day, to stand back and enjoy and appreciate what has been accomplished.

Stones for Africa! And they were all harvested from the surrounding hillside, where they had been waiting for millions of years to be re-assembled into providing some sort of shelter.

This is living! A man and his dog inside a home created from organic, "real" materials — wood and stone, products of mother earth: not man-made plastics, artificial fibers, and synthetics! (Yuk!) The large rock that I am sitting on in the doorway is the very first rock levered into place when I commenced building in December 1980. I had no drawings or plans to follow, no rules or inhibiting regulations to adhere to.

Keeping fit! I declined the offer to have a roadway bulldozed onto the site — preferring instead, to carry all building materials up the hill. This included water (prior to the installation of a pump).

Pausing to listen to the bark of a distant baboon on the other side of the valley—(Note the 2-liter Coke bottles: four of them, two in each hand, were carried up the hill for the cement mix. Fortunately, I now pump the water up the hill to a storage tank from where it "gravity feeds" back down to the house.)

more...

Excerpts from Ian's Letters

I BUILT MYSELF a simple stone house here on the side of a steep hill overlooking the naturist resort of Beau Valley . . .

Thank You! for your complimentary remarks about my stone house. It would be great if it was featured in a future edition of *Shelter*, possibly bringing me into contact with other builders of unconventional houses throughout the world . . .

Not having been restricted by any plans or drawings on paper, I have had a free rein to "feel it" as I went along—which is the best way, of course, for the expression of ideas as they come along. The end result is far more rewarding than I had hoped for.

Visitors who climbed the hill to see what was going on used to say they thought I was mad to be building in such an inaccessible place. "You're crazy!" they said. "Yeah, I guess so," I replied — "*Stone* crazy!"

Work on my stone house continues . . . the latest addition being a fully insulated, cantilevered bathroom with the bath set into the floor alongside a large picture window with a great view out over the valley below.

. . . the studio which I intend to build will be behind the house, in amongst the trees. More importantly perhaps, it will provide me with an on-going project for the next few years—I don't really want to come to the end of this to-date, 14-year project! (I am in awe of Simon Rodia's 33 years of involvement with his Watts Towers which he singlehandedly created.) . . .

Incidentally, fly-screen gauze, and not glass, was used to cover the window spaces. End result, an economical solution—which also allowed the house to "breathe." Furthermore, baboons leaning against the wire gauze merely caused it to bulge here and there, without causing any real damage. (Glass would have broken.)

The great master builder, Frank Lloyd Wright, once said that it seemed to him that no one who had any love for landscape could ever impose a rectilinear, geometric pattern upon the face of the earth, as such a pattern was fundamentally alien to nature. Bearing this in mind, the shapes, sizes, and colors that I incorporated as I went along with the creation of my simple stone dwelling were influenced by the thoughts of this organic architect.

(Ian commenting on one of the many magazine articles written about him): As you can see, they too, liked the idea of my living alone with the baboons (and leopards!) — "searching for a mate —" they said. Okay, that's true enough, for I wish I had found the right gal to share my life, and my cave, with . . . but I don't go around searching . . . she'll come along one day(?). Or will she?!

Re-reading *Shelter* for the umpteenth time is a constant source of inspiration and pleasure — "jam-packed with ideas," as they say — *great* stuff.

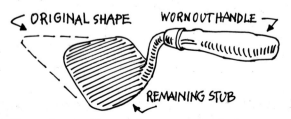

Wear and tear on 8" trowel. Over years, reduced to less than half its original size.

Kitchen viewed from passageway, as shown in bottom photograph, i.e., reverse angle. (Note: View of distant hills through kitchen windows unfortunately "burnt out" in photo.)

I (finally) . . . conceded that having electricity "made sense"— and laid a long length of heavy cable (concealed) down the hillside. This enabled me to listen to some good music whilst relaxing in the bathtub. Also, my guests got a chunk of ice in their sun-downer drinks out on the front deck. (They then forgave me for the electricity!)

This was the heaviest object carried up the hill — a hollowed-out chunk of granite, once used by Zulus when they ground their maize (corn) by hand, using a rounded stone in the laborious process. It had been given to me as a gift years before, and found its final resting place high on the hill alongside a shrub near the top of the stairway leading to "Bally High!"

A pause while I ponder my next move: Big game hunters do the leg-up bit with their prized trophy, rifle alongside — this "big-un" was a meaty challenge for a 60-year-old equipped with shovel, a lever-bar, and lots of determination. The rock now rests usefully at the entrance to my studio (as shown in photo below, up 12 more steps!).

I can step out of my newly constructed studio and onto the roof of the dwelling below. This photo: taken early October — new leaves on the trees and shrubs. As you can see, I'm usually bare-foot (and bare-bum) — I enjoy the feel of mother earth beneath my skin. Shoes or boots have been avoided ever since I started building . . . oops, apart from a few stubbed toes, no injuries to date, thankfully.

Above and below: My hill-hugging studio, dug into the rocky hillside behind the house proper. The oval-shaped room commanded a splendid panoramic vista over the Valley through the stepped windows, which were covered with fly-screen netting, thus allowing the free flow of air. When my baboon neighbors paid me a visit, not only did the troop test the strength of my roof, but they had a habit of sitting on the window ledges, often leaning against the fine wire netting, which withstood the pressure better than glass would have. The heavier-than-preferred fascia-board was eventually textured and painted a sandstone color. I built a rough stable door at the entrance.

Early view of my kitchen, but not of the splendid vista across the valley below. Recessed into the rear stone wall, a small electric fridge. The two-seater table alongside secured to the wall, and supported by a single stout pole in the middle. (Photo, Barry Comber)

This sketch illustrates the all-too-familiar scene . . . my neighbors making themselves at home, the old chap, alpha-male, posting himself as lookout atop the chimney. Baboons have one of the loudest "sound-boxes" in the animal kingdom. Their "ba-hoo!" bark, as they effortlessly scramble up and down the sheer rock faces, echoes across the valley, and for me, epitomizes "Africa."

more...

As viewed through a telephoto lens from across the valley . . . the result of hard labor: plenty-plenty Home Work! When I began putting the first few stones together, visitors who climbed up the hill said, "Jeez, I think you are crazy!" Now when they visit, they say "Gee, aren't you lucky!" I explain to them that the difference between being crazy and being lucky is 13 years of hard work — of sticking with it . . . and I tell them to ponder on that wonderful Japanese proverb which says that a journey of a thousand miles began with a single step.

BALLY HIGH

"In the bathroom, a shower has been simply constructed. At the bottom of a series of perfectly shaped stone steps lies an enclosure where one can shower in total privacy, with solid earthy rocks enshrouding your pleasurable experience. The basin has a tap that runs water, through gravity, into bowl-shaped rocks. The bathroom faces an immaculate scene which has a touch of sensuality about it. While bathing you are able to watch wildlife in its most natural form."

–Home and Garden

An additional feature of the home is a rock paddling pool set in and amongst the endless wilderness. On hot summer nights you can slowly unwind in the comfort of cool water. The art studio overwhelms with inspiration with an immense view of the surrounding mountains that can be seen from the artist's seat.

–Home and Garden

Up to roof height over the shower area — working through the dry winter months when no rain falls.

The sunken shower that I built — steps leading down to it at the side of the tree-stump towel "rack," flat rocks formed the wash hand-basin — a suitably curved stone set in the wall, my soap dish. Green ferns sprouted from in between the chunky stonework. From above, two separate sprays delivered hot or cold water, depending on which length of cord you pulled — (hot water came from coil of black plastic piping on rooftop). A skylight allowed natural lighting — diffused and softened by having painted the glass canopy white on the underside.

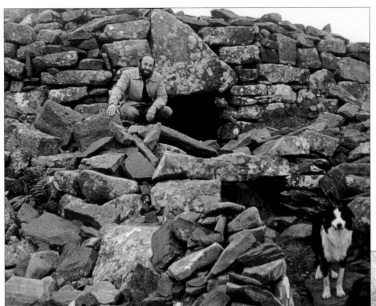

Lloyd, have you ever heard heard of Skara Brae on the Orkney Islands off the top of Scotland? No? It's a Neolithic settlement in a good state of preservation dating back some 3000 years B.C.! A cluster of stone dwellings revealing their way of life all those years ago . . . I visited the site in 1984. . . .

1987, on the west coast of Scotland, at the entrance to a neolithic "Broch." The low entrance was to ensure that anyone entering would be vulnerable to "instant dispatch" should they be unwelcome. These fine examples of stonework dating back thousands of years, have been an inspiration — as well as being a humbling experience. Needless to say, I have often wondered whether or not my own stone creation will still be standing in 2000 years' time — or merely a pile of rubble!

At a traditional "Black House" on the Isle of Skye, land of my ancestors. The Black House gained its name due to the smoke-filled dark interior, where a peat fire burned continuously. Note the stone weights hanging from the thatch — a precaution against gale-force winds.

This photograph taken at the Royal Scottish Gathering in Johannesburg some years ago. The young pipeband members wanted to pose with the replica Claymore sword and targe I had made.

"Gathering of the clans" — well, clan shields, anyway. This is my collection of targes — in Gaelic, a targaid. They were used by Highland clansmen in battle, carried on the left arm by means of leather straps.

Epilog

The land on which Ian built was sold in 1998, and he had to abandon the house. Since then he's been living at a friend's house, and looking for his next adventure.

All materials for the building are from the land. The main logs are red pine; purlins (on top of roof beams) are maple. Everything is pegged together with ash dowels. The porch cantilevers out 10 feet. The wine cellar is underneath.

Rocking chair made by a friend of Bill's, framed with willow; seat, rockers, and back are oak.

Downstairs bedroom

POLLYWOGG HOLLER WORLD HEADQUARTERS

Bill and Barb Castle

On a trip to Costa Rica in 1990, I am driving down a dark road south of Puerto Viejo (on the Caribbean coast) one night, looking for a place to stay, and see a hand-lettered sign in a clearing saying "Bed & Breakfast." I go down a dirt road to a fence, park and walk up to a building where a bunch of people are sitting around a table by lantern-light, talking and drinking beer.

Hosts are Bill and Barb Castle, from the Alleghany Mountains in southeast New York State, and they had rented ten acres of Caribbean beachfront land on the edge of the jungle to run their B&B and to explore Costa Rica. Bill and I hit it off from the start. He's an ex-general contractor who had specialized in heavy construction — a genuine builder. He has energy, a sense of humor, and a sense of adventure. The room we are in that night turns out to be the kitchen, the bottom story, with a (packed) dirt floor, of a two-story pole structure that Bill had put up in two days, with help from Barb and his son Quentin. There are four corner poles, the ground level kitchen, and a sleeping loft you reach by a ladder. The loft is Bill and Barb's bedroom, with open walls to catch the ocean breezes and a roof thatched with palm fronds from nearby trees. The little building is competent, practical, made of local materials, tuned into the climate, and it works. Naturally Bill and I talk about building. He shows me a picture of the log home he had built in the Alleghanies. Wow! Now *here* is a log cabin. I decide right there I'll visit the Castle compound (Pollywogg Holler) before long.

About six months later I go to the American Booksellers Convention in NYC, and afterwards get on a plane to Buffalo, where Bill picks me up at the airport.

When we get to Bill's property, we park by a studio/workshop building that is hooked up to electric lines. It's Bill's office, workshop, and son Mick's photo lab. To get to the house we walk down a graceful path a half-mile through the leafy woods. Here and there are sculptures, a hand-hewn lovers' bench, finally a bridge going across a creek. There's a stunning log building on the side of a hill, looking down on a pond. Next to the house is a perfect little sauna building, like something you might have seen 100 years ago in Norway or Russia, crafted by a man with an axe. Bill is one of the builders that I consider at the heart of this book. They are guys that are doing such unique work that I have wanted, for a long time, to tell people about them.

Bill and Barb run a very together and very rural bed and breakfast facility. Barb makes healthy, tasty food on a woodstove, Bill serves his own champagne (they make a few hundred bottles a year). Guests use the wood-fired sauna, then sleep in a cozy loft on the sauna building's second floor (with balcony deck outside). There's a 20'-deep artesian well lined with bricks.

I sleep on the deck of the house; it's like sleeping in the woods, but better. The bed is comfortable, with homey blankets, the trees fragrant. I like this place!

Bill relaxing in the main room. Note the wagon wheel kerosene-light chandelier, which can be raised or lowered. (It's counterbalanced.)

Wood cookstove in kitchen

more…

The sauna is about 100 feet from the house. Upper level is the bedroom for weekend visitors.

Entrance porch to sauna; the other side of the building from the view shown above

THE NEXT DAY Barb's friend Brandi comes over and they do their own aerobics class out by the pond to an exercise tape. Barb's a modern woman, but with the skills and strength of a homesteader. There are pictures of her shoveling concrete and handling logs when the place was under construction. "She mixed 90% of the mortar," says Bill. This is a dynamite duo. A couple arrive to stay for the night, and we hang out, walk in the woods; I shoot lots of photos. The camera loves it here! That afternoon, Bill is bottling champagne in the open-air cellar. Which is where we all gravitate to. We sample . . . and sample . . . hey, this is the good part of booze. Bacchus smiles on us that day.

On these pages is the story of the Castle family and the charming, tuned-in retreat they have created in the Alleghany woods.

Pollywogg Holler is now a fairly well-known eco-resort. People hike to the house, enjoy Barb's home-cooked meals and Bill's champagne, take saunas, relax, meditate, and wander around through trails in the woods. Nearby are horse rentals, swimming in the summer, and downhill and cross-country skiing in the winter. For info, call 800-291-9668.

Interior of sauna, with weekend visitor looking down from bedroom. Door shown goes into "Hot Room." Wood fireplace on left.

Door to sauna, where Bill has carved "Bathe Often, Never Hurry" — inspired by a Grateful Dead poster

Rustic guest shed in the woods

Outhouse

Above and left: Details Bill has carved on log ends of sauna building

))) www.pollywoggholler.com

more...

Kitchen, and Barb preparing the evening's meal on wood cookstove

Bill and Barb Castle bought their land in 1976, 25 acres in the Allegheny Mountains bordered by state land, with a small stream running through it.

Bill: "It was the centennial year, and I liked that, the spirit of independence. I was also reading stuff like the *Mother Earth News*, about people building their own houses in the woods."

The Castle family, including their three kids, Mickey (14), Debbie (12), and Quentin (9), started work on the property on weekends. The building site they chose was about a mile from the nearest road. "We dedicated all our spare time to the place for about two years."

Guest sleeping porch

Mickey

Building in the '70s

Barb levering log onto wagon. Note the mud. "It seemed like it rained every weekend that summer."

First log notched, ready to be rolled back over onto sill log

Bill and Quentin hauling a purlin into place with gin pole

Quentin (9) turning crank on gin pole

Bill leveling off top of purlin

Moving log with tongs

At work with adze

A friend at work

Bill shown at left, building his brick-lined well. Note spiral of projecting bricks, which Bill set in place so as to have a stairway going down the well.

Below left and below: In later years, Bill framed a roof over the well, which he and friends thatched with straw.

BILL MADE A DEAL with the state to cut logs for the house on adjacent forest land. "We'd go in and mark the trees; they were red pine. 90 trees cost me about $45 total. I cut 'em, skidded them out to the edge and took the bark off. The longest was 35′. I had a little old (1953) Allis Chalmers model C 18-horse tractor."

They built the foundation out of large rocks, some weighing a half-ton each. Bill's heavy construction experience came in handy. "I know how to move big things." Bill constructed a gin pole to move the logs into place; the heaviest was about 1400 pounds. "I'd roll a log into place and mark it; then I'd roll it back 180° and work on the surface. I cut the notches by eye." Once they got the walls up, Bill wasn't sure how to frame the roof. "I kept thinkin' about it. I was goin' over two or three things in my mind and one Sunday we went to church. I was settin' there looking at the ceiling and here's this magnificent scissors truss. I couldn't wait to get outta there." He hurried home and got started on the roof that Sunday. "I put a pole at either end of the building and stretched a string across. Then I built a scaffold and brought the peak of each truss up to that height."

Bill used a Homelite 150 chain saw (no electricity at the site, so no motor-driven tools) for the cuts. Also a little coopers adze he got at an auction ("my favorite tool"). No nails were used in the log framing. Bill cut pieces of ash in the woods and split out the dowels. (They're called trunnels, a word derived from "tree nails.")

After the trusses were in place, Bill pegged the maple purlins in place *(see construction photos above)*. Then he flattened the tops of each purlin with a chain saw and an Alaskan mill attachment. 1″ × 10″ boards were nailed onto the purlins, then a layer of thick 10-mil black polyethylene on top of that, then another layer of boards on top of the poly, nailed only at the peak and ends (so there were no holes in the poly). Then pieces of rustic slab wood over that, and they had a watertight roof.

For the cracks between the logs in the walls, Bill used thin (2½″) strips of metal mesh (attached with galvanized nails), plastered with mortar — on the inside and outside — with fiberglass insulation in between. Barb mixed all the mortar in a wheelbarrow with a hoe.

For flooring they tore down an old silo, and used the clear cypress tongue-and-groove boards.

In 1980 they moved in, and have lived there ever since. "I built the place as a hunting cabin. I never dreamed we'd be livin' in it."

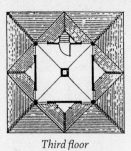

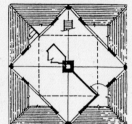
Third floor

Second floor

First floor

SEARCH FOR RADIANCE

John Silverio

John Silverio is an architect who designed and lives in this lovely radial-patterned house in Maine. He was influenced by Norwegian stave churches and inspired by his concept of ". . . designing radiant forms according to spiritual principles." Shown at right is a quilt by Susan Silverio that shows square spiraling, as do the three floor plans of the house above it. Below are excerpts from a paper by John titled: A Search for Radiance: An Architect's Credo.

IDEAS ABOUT HOW form takes on radiance lead us to ask how the materials used in construction affect the quality of radiance. To maximize the flow of energy through the shell of a building, as through a natural body, the shell should be made from natural materials which are harmonious with the environment. The reds of bricks and grays of stone and rough wood take on iridescence, the patina of weathering, as they age. The inner quality of a building constructed of whole natural materials shines through as it becomes a part of the surrounding environment.

Nature, being the source of all building materials, is also the reclaimer of all materials. I envision life and death cycles of architecture. A building should be a living part of these cycles rather than a foreign object transplanted into nature. Architecture is thus seen as a materializing process or, in other words, a birth process. Nature offers the material and man provides the energy.

Natural materials should be transformed as little as possible. Wood and stone should be left rough and unfinished since rough, porous, and weathered surfaces seem to catch and hold particles of sunlight, producing a warm and radiant glow. The use of tools upon materials should leave their record so that any necessary changes are revealed in the final product.

All transforming processes such as aging, heating, and shaping should be kept to a minimum. If transforming processes are complex, such as conversion of raw materials into plastics, the end products no longer reveal their sources. What we see in nature does not prepare us to look at surfaces and comprehend materials which are man-made. No material on earth is entirely man-made; humans simply isolate and transform substances from nature. We feel most comfortable when surrounded by natural materials.

As for transporting materials, I favor the use of materials from the locale. Foreign materials, out of context, are disorienting. Local materials are time-tested to climate, adapted to use, and compatible together. If materials for building are gathered from the vicinity, the energy involved is more intensified than if lines of transportation are stretched out.

Not only should materials be found nearby, but I favor using them as they exist in the landscape. Stone, found close to the ground and only slightly visible, could be used in foundations. Wood, found above ground and plentiful, could be used for the upper structure.

A building built with such considerations for materials should have a truly healthful, vibrant and radiant quality. We might sense an aura

around it. Lacking the deadness of concrete or the reflectiveness of metals, such architecture actually breathes and merges with the whole vibrance of life.

Forces Molding Form One of the keys to understanding how radiance manifests, either in a person or in architecture, is to visualize the human body or the walls of buildings as hollow shells. These shells become enlivened as the forces of life, within and without, move through them. When a great deal of spiritual energy moves through a shell, it is not only alive but radiant with life.

With architecture there are two primary directions of forces, the outgoing/expanding and the incoming/contracting. The outgoing/expanding forces push out on the shell and produce bulges in the form. These outgoing/expanding forces are the occupants, their possessions, their actions or movements, their thoughts and even their prayers. These are expressions emitted outwardly into the universe and they tend to expand the spaces of living. The bay window is a manifestation of such outgoing/expanding forces.

The incoming/contracting forces, conversely, push in on the shell producing recesses or pockets. These incoming/contracting forces are nature, neighboring society, and universal forces in a more spiritual form such as grace and intuition. These are impressions focusing inwardly from the universe and they tend to contract the spaces of living. The recessed entrance is an example of such incoming/contracting forces. Through these two force directions, which produce convex and concave shapes, we are able to distinguish elements in the world of form.

As an architect I seek to create a shell which resolves the forces acting upon it, a shape of composure and balance, one which is centered. The shell need not be a barrier but can actually amplify the flow of forces when it balances concave and convex shapes.

Exterior walls are built of "rastra block," an efficient building material consisting of 85% recycled polystyrene waste.

HOUSE ON THE EDGE OF TIME

Paul Nonnast

Bathroom

PAUL NONNAST is an artist living in Jerome, Arizona, who designs and builds houses. His shop (below left) is built of stone, concrete, steel and glass. His house, shown in the other photos, is built of rastra block. He likes using "non-precious" building materials which, he says, ". . . if used right, can be elegant." For example poured-in-place concrete walls using rough-textured wood forms, or the cornice detail shown below where he used plain galvanized metal trim, but carefully mitered instead of cut at right angles. He favors designs that ". . . don't require a lot of space to make a spacious house."

For building materials, Paul says he uses ". . . any junk I can find that has use."

Mitered cornice detail

Rough-textured, poured-in-place concrete walls

Inspired by John, Ian Ingersoll built this mortise and tenon house from old barn beams. That's (left-to-right), Ian, Caleb, and John. The house later burned down and Ian built another; he also went on to found a thriving Shaker furniture-making company in West Cornwall, Connecticut.

JOHN WELLES

In 1970 I WAS BUILDING geodesic domes at a hippie high school in the Santa Cruz (California) mountains. J. Baldwin, who had worked with Bucky Fuller, and I had been hired as teachers and were helping the students build their own living quarters while at the same time experimenting with different building materials and techniques.

One day a chopped green VW bug pulled up outside our workshop, and out stepped a fast-talking, pipe-smoking big man, clad in overalls, with a twinkle in his eye. John Welles, from Connecticut, was visiting "alternative" builders and home energy producers on the West Coast. He was a competent jack-of-all-trades, an inventor, builder, gardener, welder, and sailor with an inquiring mind. He'd built his own house in the Connecticut woods out of old barn timbers. He sprayed polyurethane foam for a while. He knew how to build mortise-and-tenon barn frameworks. He had a backhoe, understood how to move heavy objects around, made windmills and solar water heaters, and redesigned and rebuilt a variety of motorized vehicles.

John and I have kept in touch for about 35 years now. Every year or two he spends a few days with us in California, and I've been to visit him in Connecticut several times over the years. Eventually we drive each other crazy with our West Coast vs. East Coast sensibilities, but not before we have run through our latest excitements in building and gardening and alternative energy.

The photo at left was taken in November in the '80s. My family and I were visiting John for a few days. This day he fitted out our two boys, aged 10 and 12, with welding masks and they were applying John's welding torch to rocks to see the sparks fly (boys will be boys!). He then showed us the little 4-wheel-drive, 4-wheel-steering dump truck he had assembled for driving into the woods to get firewood. The rear wheels were steered with a hydraulic valve. It could literally turn on a dime. He built the dump body out of scrap steel and used an old dump truck lifting mechanism. The frame is two front sections, one facing forward, the other backward. It has heavy springs and can be loaded with as much cargo as will fit into it.

John using the jack at an angle to push the building along

He had two sets of 4×8s, and would move one around to the front when he got to the end of the other.

House Moving Solo

In the early '80s, John bought a small piece of land down the road from me in Bolinas. I had built a small, funky Japanese-inspired building that I wanted to get rid of. I think we agreed on $500. One day John said he was going to move the building down to his land (about ¼ mile) the next day. "How?" I asked. With a 48″ bumper jack, he said, and he would roll it on logs. Not only that, but he was going to do it solo. Sure, John.

I had to go to Berkeley the next day, and I didn't get home until the afternoon. I went out to look at the building, and lo and behold, it was halfway through my gate out in the road. I shot these photos of John and his technique.

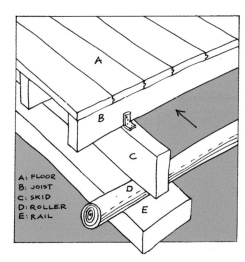

A: FLOOR
B: JOIST
C: SKID
D: ROLLER
E: RAIL

John, pipe puffing, pulling 4×8

By the end of the day he was 500' down the road.

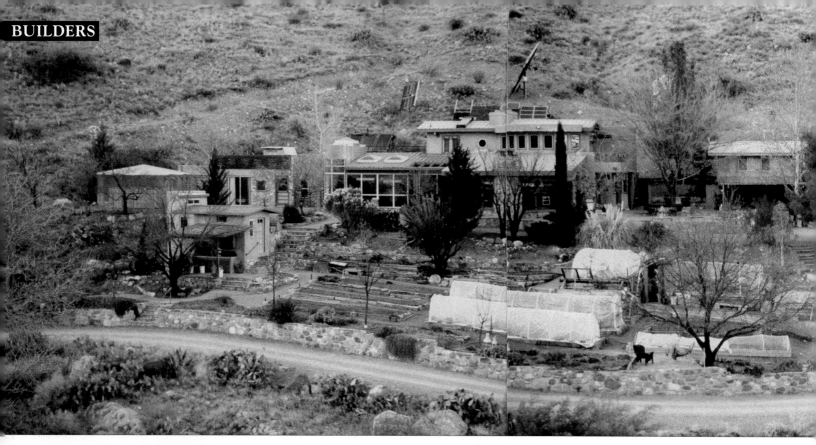

HIGH DESERT HOMESTEAD

Indoor greenhouse provides food as well as moist air.

HERE, ON THE BANKS of an arroyo in Arizona (near the old mining town of Jerome) is a unique solar-powered homestead, tuned into the land and intelligent in design: a perfect homestead, both beautiful and functional.

Rennie Radoccia and Andrea McShane Radoccia met at Paolo Soleri's Arcosanti building project in 1974. They lived in tipis for years and in fact built their first house around a tipi, then took it down when the house was ready for a roof. Their present homestead consists of their home and outbuildings built along one side of the

arroyo, with Rennie's architectural studio on the other side. For the house they used local rock for the lower walls, and studs for the upper stories; the buildings are insulated with both fiberglass and blown-in cellulose. They made all the furniture in the house and Andrea made the kitchen tiles. In addition to being a potter, she runs a belly dance troupe and is a certified massage practitioner. Rennie is a rarity among architects: one who can build with his own hands, and whose designs are practical as well as aesthetically pleasing. Together they make a great team. Says Rennie, "Building is our passion."

The "hoop greenhouses" shown above use bent-over ½" PVC pipe on 2½'-long re-bar stakes and 6-mil, 4-year polyethylene from Arizona Bag Co. (602-272-1333). There is a wire ridge beam and zippers (www.charleysgreenhouse.com) to keep heat in at night. They were full of vegetables in January when we visited, when the nights were freezing.

Rennie's 1500-sq.-ft. studio is built of "rastra block," a highly efficient and ecological building material, 85% of which consists of recycled polystyrene waste. The blocks are $10' \times 15'' \times 10''$, weigh 135 lbs. each and are epoxied together; then the 15"

o.c. cells are filled with concrete and re-bar reinforcement. Rennie calls it an "owner-builder-friendly product" (www.rastra.com).

Their solar energy system consists of 36 100-watt Siemens solar panels on Zomework trackers, a 4 kw inverter, 24-volt, 20-year lead/acid batteries, and a Kohler propane backup generator. The system can run the washing machine, kitchen stuff and a Skilsaw. If you look closely at the right of the large photo above, the yellow object is Rennie's miniature Caterpillar bulldozer.

Kitchen counter tiles made by Andrea

Andrea's belly dancing outfits

THE YURTS OF BILL COPERTHWAITE

BILL COPERTHWAITE doesn't have email, doesn't have a phone, and lives in the Maine woods a few miles from the nearest roads. When I visited him in the '70s I walked in a mile or so through the woods. You can also get there by canoe down the coast. My son Peter was with me and we spent a few days there, taking canoe trips in the inlets, and hanging out with Bill and his apprentices. Bill has a Ph.D. in education from Harvard, worked for two years in Mexico with the American Friends Service Committee, designed a traveling museum of Eskimo culture, and has lectured all over the world.

In 1962, while reading a *National Geographic* article, Bill recognized the folk genius in the design of the traditional Mongolian yurt. He found in the

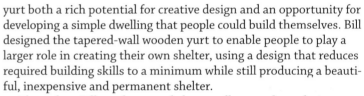

yurt both a rich potential for creative design and an opportunity for developing a simple dwelling that people could build themselves. Bill designed the tapered-wall wooden yurt to enable people to play a larger role in creating their own shelter, using a design that reduces required building skills to a minimum while still producing a beautiful, inexpensive and permanent shelter.

These days Bill conducts workshops, sells yurt plans, designs and consults on yurt projects, and continues his search for ways to simplify life in the 21st century. Chelsea Green has just published Bill's *A Handmade Life — In Search of Simplicity*. To contact Bill, and for web information on his Yurt Foundation, see the next page.

The yurt shown in the three photos on this page is Bill's home in the Maine woods. It is 54' (eaves) in diameter and was designed so it could be built over a period of several years and still provide shelter during the process. It is a tricentric, or three-ring yurt with 2700 sq. ft. of floor space. You can first build the

16' inner core as a room to move into. In the second stage, you can build the large sheltering roof over a gravel pad, allowing the major cost, floor construction, to be delayed. In the meantime you have a spacious area under roof that can be used for a workshop, greenhouse, garage, or for play.

The standard yurt can be built at 17' (eaves) diameter (and also at 12' and 10'). This is the simplest to build, makes a great cabin for one, or seminar space for 15 people, and can be used as a summer camp or mountain retreat. A circular skylight spreads illumination evenly, and a ring of soft peripheral light enters though the windows under the eaves. People have used these as saunas, guest rooms, and as offices with curving desks.

33' freespan yurt at The Mountain Institute, Cherry Grove, West Virginia, 1991

Guest yurt at The Yurt Foundation, Dickinsons Reach, Maine, 1966

Concentric yurt on Mother Earth News *land in North Carolina, 1979*

The concentric yurt is 38' (eaves) diameter and is really one yurt inside another. The inner yurt supports the roof of the outer one and reduces materials costs. This concentric way of dividing a circle creates a unique free-flowing space in the outer ring and a secluded feeling in the inner loft yurt. Since the inner yurt is raised a full story, it provides a room underneath that can be used as a bathroom, storage room, pantry, or living room. These yurts have been used all over America as permanent homes, summer homes, and common rooms in communities. It has 1000 sq. ft. of floor space.

Helen Nearing's yurt at Harborside, Maine, 1990

First 54' tricentric yurt, at The Mountain Institute, Cherry Grove, W.V., 1976

Inside a Travel Study Community School yurt in Franklin, N.H., 1968

Plans for the 3 basic yurts shown on these pages are $25, $50, and $75.

CONTACT: The Yurt Foundation, Dickinsons Reach, Machiasport, ME 04655

WEB INFO on workshops, yurt plans, yurt calendars:
http://www.yurtsource.com/yurtfoundation.php

WEB: yurt photos (hundreds of them):
Go to: http://www.altavista.com, click on "images," and type in "yurt."

more…

*Some comments by William Coperthwaite
on his philosophy, background, and work*

IT IS REPORTED that I was born in Maine in 1930. For the past 43 years, I have lived in the wilderness on the coast of Maine, seeking to discover simpler and more elegant ways of living that can be of use in building a saner society. This has been my base as I have studied, traveled, and lectured around the world in search of ideas as to how we might live fuller and less exploitative lives. We have a tremendous potential to design a better society, could we but learn to tap that potential. I've come to believe that we can blend the best of age-old folk wisdom with the best of modern knowledge to create a world of incredible beauty — a world of caring, creativity, and joy.

A society that aims for the happiness and fulfillment of the largest possible number of people, and which is concerned for the ecological balance of the planet, will see the necessity, the beauty and the wisdom of living simply. My adult life has been spent seeking out cultures and individuals who have something special to contribute to simpler, more healthful living — be it in the area of child care, gardening, community planning, handcrafts, structures, or design.

My work in designing the modern yurt grew out of this research. The design derives from the blending of the folk genius of inner Asia with modern materials for a structure that is light and strong, inexpensive and easy to build. Since 1964, I have designed and guided the building of some 300 yurts from Alaska to Florida, from Maine to California and in a number of other countries. They vary in size from small play yurts to four-story ones 60 feet in diameter. They are in use as homes, classrooms, mountain retreats, summer homes, and saunas. Using the modern yurt as a symbol of cultural blending, I set up the non-profit Yurt Foundation in 1971 to collect knowledge of simple living around the world.

HOMES

"A house is a home when it shelters the body and comforts the soul." –*Phillip Moffitt*

JACK WILLIAMS

JACK WILLIAMS built a house in the Northern California woods; it's a dream homestead, built with imagination, integrity, and sweat. The house faces south, looking down on three miles of forested land to the blue Pacific Ocean.

He cut redwood saplings on his property (". . . more sapwood than heartwood") for poles. He poured piers 4' on center, connected by a grade beam, and attached the poles to the piers with metal straps. From the ground up to about 24" he built a ferro-cement wall, using about six layers of chicken wire (on the outside of the poles), plastered with sand and concrete. He says if he had it to do over, he'd use expanded metal lath instead of the chicken wire. The poles have held up well, he says, since they're protected from the weather.

Jack was one of the Northern California "off-the-grid" pioneers. For some 20 years now, he's had his electricity coming from 16 solar panels, four of which are devoted to pumping water from a well. He uses a 2000-watt Trace inverter and has three 500-amp-hour forklift batteries (which he bought 16 years ago). For backup during winter months he has a 6500-watt propane generator. He stays in touch with the rest of the world with a cellular phone and a TV satellite dish. Jack has fruit trees and grows vegetables, and these days he's working on a new building with a ferro-cement roof.

Front of house

Rear of house (interior of this wall shown in photo at middle right)

Jack's solar convection fruit dryer. Trays inside black box hold fruit.

Jack bringing trays of dried fruit into the house

Bathtub in adjacent greenhouse   *Marley*

Above and below: The house that Kate built

Kate built this couch out of local alder and willow, using nails and grabbers.

KATE TODD

The house Kate now lives in, showing solar collector for outdoor shower and photovoltaic solar panels that charge batteries

Upstairs bedroom of house at lower left

Kitchen

KATE TODD built two off-the-grid houses in the Northern California woods in the early '70s. She and her partner started the house (shown in the two photos at top of left page) in Spring of 1972 and moved in that winter (with one wall covered with plastic sheeting). The foundation was concrete piers. The pole frame was spiked to the piers with a vertical piece of 1″ galvanized pipe. For the attached greenhouse she poured a perimeter foundation. "I had help from a lot of friends."

Three years later she built the second house (other photos on these two pages) by herself. When her two kids got into their teens, she let them have the second house to themselves. Kate's place is a cozy little wooden house with good vibes and although it may be tiny, it's a real *home*. Both houses have electricity produced in the winter by a Harris-Burkhardt hydroelectric generator, a small-scale pelton wheel powered by water from a 1″ pipe directed from an uphill creek. In the summer, electricity is provided

by photovoltaic panels. Both systems charge batteries and Kate runs lights, a coffee grinder, radio, a tape deck and once a week, a vacuum cleaner, the sewing machine and/or a VCR. "The great thing about hydro is it's 24-hours a day." A little electric heater goes on to take any overflow of electricity from the hydro system and avoid overcharging the batteries. Hot water comes from a "Blazing Showers" woodstove coil in winter, and a solar collector for the outdoor shower in summer. Kate also has a productive garden and is a print-maker. She teaches English as a Second Language and travels whenever she can — Nepal, Bali, Italy, Mexico, Cuba, Guatemala . . . . In 1986 she took a hydroelectric generator to Nicaragua and helped set it up to provide an electric light to each of nine houses in a small village cooperative. She drives a 1993 Nissan pickup truck and she and her partner just bought a Toyota Prius hybrid electric car, which gets 55 miles per gallon.

In the garden

Kitchen

35

SUSAN LEWIS

Home Builder
Winemaker

I<small>N</small> 1974, S<small>USAN</small> L<small>EWIS AND</small> R<small>OSEMARY</small> W<small>ARD</small> built a solar-powered wooden frame house in the hills of Mendocino County, California, using no power tools. "Neither one of us had built anything before but a wobbly bookcase," she says. Susan was inspired by another woman homebuilder, Kate Todd *(see p. 34–35)*. There's a solar panel on the roof that runs lights, a DC refrigerator, TV/VCR, hair dryer, and washing machine. There's a Holly hot water heater unit in the woodstove, an outdoor shower, and a composting toilet. The 120′ well is pumped by a 12V DC pump and two solar panels.

There are five acres planted in chardonnay grapes; Susan makes wine and champagne and sells surplus grapes to wineries. She has a '92 John Deere 1070 tractor and an immaculate 1953 Chevy pickup truck. She has 12 chickens, five cats, and a Chesapeake Bay retriever.

JOHN FOX

JOHN FOX bought 40 acres of forested, steep land in Northern California in 1970 and built his hand-crafted house bit by bit. It's remote: the road ends about 500 feet above the house, and John has a 470' ⅜" cable and a winch that he uses for hauling groceries and supplies to the house. He has gravity-flow water from a creek that powers a Water Watts microhydro turbine for electric power during the wet season, and solar panels for the rest of the year. There are four L-16 deep-cell batteries for storage and a Trace Inverter that converts the DC to AC. There's a Honda generator for backup. The house consists of two seven-sided sections and is dug into the hill. It's light, airy, and colorful, and has the feeling of a treehouse. In the last four years, John's son Heron (shown on the rope swing below, left) has been working with John on construction and gardening.

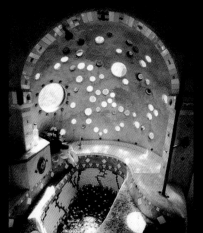

John's vintage copy of Shelter

ON THE BEACH

I**N THE LATE** '60s, Karen and Roger Knoebber and their three young children lived for about a year in a driftwood house on a deserted beach north of San Francisco.

LLOYD: How did you end up doing this?

K**AREN**: We left Berkeley around 1967 and went on the road in a camper. We got as far as Key West and ran out of money. There was a little island called Boca Chita next to Key West that was an abandoned Navy base. It was deserted, so we hauled lumber across from Key West and built a beautiful little breezy (*meaning no walls*) house. Friends would come out and say, "What a way to live!"

They left there after about six months, moved into an abandoned farm house in Maryland for a while, then headed back to California. Once again they were broke, with no money to pay rent.

Above right: Karen
Above: Karen with Shufina, Khamoor, and Cosmo
Left: Roger with Shufina, Khamoor, and Cosmo

Looking back, it's hard to believe you could ever do something like this, just an hour away from San Francisco. A home that costs practically nothing. No taxes, building inspectors, electricity, cars, roads. Or look at the photos of '60s New Mexico communes on the next two pages. Are there things like this going on in America now? Could this be the same planet?

The sauna

"We heard about this little driftwood community on the beach, and we went out there. There were about eight houses, and they let us live in a tiny little place while we ran up and down the beach collecting driftwood for our house."

How old were the kids?
Five, four and two. They were all born in San Francisco.

How old were you at the time?
28.

What were your days like at the beach?
We'd walk a quarter-mile to get drinking water and we'd walk along the beach picking up firewood. The kids would play on the beach.

How did you cook?
On a Coleman stove.

How about food?
We ate a lot of mussels, Roger caught fish, there was New Zealand spinach growing nearby. We'd walk into town (about two miles) every two weeks for supplies.

Why did you leave?
We got a little money, and our daughters were of school age. Also, the house was perched on shaky cliffs and we worried that it might collapse. And the authorities started showing up and telling people they'd have to leave. They said the place was too notorious.

Isn't it amazing that you could do something like that?
Yeah, nowadays you'd get arrested right away. You know, the times were just so different.

Where did you go?
We moved to Mendocino. It seems we were constantly building. We lived without electricity or a refrigerator for about ten years.

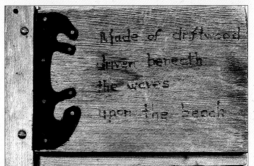

Made of driftwood
driven beneath
the waves
upon the beach

All these photos are from Karen's album, which (naturally) has a driftwood cover.

Footnote: By the time the family left, their house was the only one left. All the others had been destroyed either by their owners or the authorities.

Update: Karen now lives in Mendocino County in a house she built herself. Roger has been living in Paris for 20 years. Karen's three kids all live in California and she has six grandchildren. Karen's kids sometimes tell her they would have liked "a little more structure" in their lives.

Cabin in Mora woods

Five Star Commune in Mora

Five Star Commune

THE NEW SETTLERS OF NEW MEXICO

Irwin Klein

During the cultural revolution of the '60s, many young people with inquiring minds and adventurous spirit set out to create new lives in rural areas of America. New Mexico, with its open spaces, cheap land, and sparse population, drew thousands of new settlers. Placitas, Morning Star, New Buffalo, Reality Construction Company, the Lama Foundation, they seem almost unreal looking back 35 years. It was a time of optimism, faith, and yes — drugs — but also a lot of hard work building and repairing adobe houses, raising children, tending animals, and living communally in the psychedelic years.

Irwin Klein was a photographer from New York who shot black and white photos with a Leica during five visits of about three months each to New Mexico from 1966–71. He was working on a book he called The New Settlers of New Mexico. *Irwin died a tragic death in 1974, not coincidentally at a time when the innocence and freedom of the earlier hippie years seemed to have dissipated, and big-city hard drugs and criminal elements had moved in. In the fall of 2002, we were contacted by his brother Alan, who had all Irwin's photos and was (is) looking for a publisher, but more importantly, wanted to share his brother's photographic vision with others. Here are excerpts from the introduction to Irwin's book, along with his beautiful photos. This will bring tears to the eyes of many who were there in those years, a time before the harsh realities of life intruded on youthful idealism and gentle optimism.*

> http://homepage.mac.com/pardass/
> IRWINKLEIN/INDEX.html
> (all one line; capitalize as above.)

THOUGH SOME photographs were shot on communes, most of them are of people living alone, in couples, families, or small groups in the little Spanish-American towns in the back country. It is sometimes hard to distinguish between a group of friends who share certain resources and spend a lot of time together and a commune, but I think that a commune has to have a sense of consciously shared responsibilities and probably, a certain formal structure. Most of my subjects live in what I would call settlements rather than communes.

Many of these people are children of the urban middle class who have abandoned the drug ghettoes of large cities, though some come from rural backgrounds. There are dropouts from the universities and relatively "straight" walks of life and a few old beatniks. As I explored the evolving situations, certain patterns and themes unfolded. There seemed to be a rite of passage from innocence to experience, and a development away from the image of the hippie toward older American archetypes like the pioneer and the independent yeoman farmer.

Some might look upon this as just a photo collection of hippies. While it's true that the pictures reflect the style and decor of a particular moment which is already passing, what interested me more was that the adventure I depict is part of a timeless movement, the perennial attempt of human beings to renew the pattern of their lives. My subjects are trying, with varying degrees of seriousness, to develop a viable way of life outside our urban technological complex, drawing whatever resources they can muster from our common past and disintegrating culture.

My own role was as much that of a participant as an observer. I came to New Mexico with much the same motives as the people I photographed. In almost every case a certain bond of friendship or intimacy was established before I began working. *The New Settlers* is part family album, part document, and part myth. I consider it as much a collective expression as my own work.

Five Star Commune

Alan, Fly and Mickey in Vallecitos

Rufus in front of church at Vallecitos

Sandy in her kitchen in El Rito

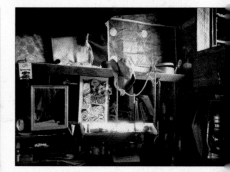

Peter Van Dresser's cabin in Potrero Canyon

Vallecitos kids

Wedding at New Buffalo

Alan & Mickey at Petreo Canyon

New Buffalo

Five Star Commune

Moe and girlfriend at Las Tables

Donna Lyons and Alana at the Seattle Gang house in El Rito

France

FUNKY

California

Nevada

"Everything I do gon' be funky from now on..."
                              -Dr. John

Northern California

A small house in the woods in the Pyrenees

ARCHILIBRE
Countercultural Builders in France

A FEW YEARS AGO we ran across a great website full of owner-built homes in the French Pyrenees. It looks as if the French have picked up on the spirit of American countercultural builders of the '60s and '70s, and the results are intriguing, especially in their work on zomes. *Vive la France!* On these ten pages are photos taken by webmeister jean soum.

http://archilibre.free.fr

Arnold's hut in the forest with walls of earth and lime and windows from junked cars

Jeanne-Marie built this pretty little house in the Pyrenees. She based the design on the old stone barns of the region, but used wood rather than stone.

Jean-Claude built his hut with recycled wood and windows.

more...

Barn of Miguel's horse

Pierrot's hut in the woods

Small house in the woods. A zome structure is attached to the other side.

Owner-built houses in the hills with view of snowy Pyrenees

Interior of large cabin

Octagonal house with slate roof

All photos on this page show Roland and Gigi's house.

THE ELVES and all the elementary beings have always known that curves and circular shapes reflect the harmony of all things. I built a round house in the heart of the forest, on a base of pink marble, surrounded by twisted and hunchbacked trees — such was my contract with the spirits to protect the harmony of this place.

–Roland

http://archilibre.free.fr

more...

A small double zome used for meditation

Zome in mountains is roofed with 100-year-old slate shingles obtained from abandoned buildings in the area.

IN THE MID-'60s, Steve Baer, a mathematician and inventor living in Albuquerque, NM, designed a series of buildings he called "zomes." Early ones were built at Drop City, the hippie commune in Farasita, Colorado, and Placitas, New Mexico. Baer published *The Dome Cookbook* in 1967, outlining the mathematics and construction of zomes. It was a wonderful, spontaneous publication that sold for $1 and inspired both Stewart Brand and his *Whole Earth Catalog*, as well as my own venture into the publishing world with *Domebook One* and *Domebook 2*.

Flash forward 35 years to Europe and you discover a bunch of zomes in the Pyrenees with a French twist. Zomes were introduced in France some 20 years ago by jean soum, who lives and works in a zome, and who sent us these photos.

soum and the zome builders added their own interpretations, and formed "Groupe Zomes" to share and exchange information. Hundreds of zomes have been built in the French countryside. They are being used for homes, meditation temples, and meeting halls.

Zome inhabitants report they are seduced by the harmony of these structures and the serenity and energy produced by the shapes. They report using " . . . small models in glass or rock crystal to increase the energetic potential of the spaces and to harmonize the vibrations of man with cosmos."

Zome with diamond motif in meadow

Fish-eye view of ceiling

jean soum's solar-heated zome office shown in photo at left and two below

Interior shows use of different materials: adobe, wood, cordwood, cob; it's insulated with sheep wool, straw, and clay.

North face; main rhombitriacontahedron is clustered with offshoots of the same geometry, producing wave-like effect.

Jean-Michel putting on the last slate shingles

http://zan.zoom.free.fr/zome_planet/z8_en.html
http://archilibre.free.fr/free/causse/zom.html
http://www.zomeworks.com
http://www.zometool.com

Looking up at interior structure of zome shown in background photo

Double zome on mountainside

Zome built on ruins of a barn

Robinson's workshop; he's a carpenter and the large zome gives him space for assembling zome components.

more…

Miguel's cabin with used glass and windows

*Bedroom ceiling
of house in
photo below*

Roof framing of circular cabin

Roof thatched with ferns, visible inside

*Hand-made solar-heated house
on south-facing slope of mountain*

Roof covered with fern thatch

These three photos show stages of building an Arigean yurt for a music studio.

Above: Locust and chestnut branches on posts, hazel branches woven horizontally

Small yurt in meadow

After plastering walls of above yurt with earth and installing windows, let the music begin!

Two different views of yurt with central fireplace

HOUSE ON THE ROCKS

Peter Marchand   *Photos by Jay Dusard*

I NEVER EXPECTED anyone to take my house seriously. It was just a quick fix at a transitional time in my life, a maverick dwelling that I put together with reworked materials and an overworked imagination. But soon after I started building I knew I had something different, something abiding. The project drew inquisitive, contemplative looks from visitors; people with far more house than mine were asking questions. In 700 square feet of shelter built on the rocks, I rediscovered simple, long-forgotten truths.

My house is located in a remote corner of Navajo County, Arizona. This is canyon country, a land of jagged contours and soft, muted colors, where layer upon layer of eroded rock creates a labyrinth of ravines and jumbled boulders. It is a wild and fanciful landscape, with weathered pines and junipers growing out of the shallow soil like forgotten bonsai in groping, conciliatory shapes, toughened by two centuries of wind and little rain.

Before I began building, I spent many days walking the land, getting to know the trees, the cliff rose, the yucca, all the subtleties of that dwarfed and windblown pinion-juniper woodland clinging to the bare bones of the high desert. But it was the rock that was so alluring— so smoothly weathered, so imperturbable, so quieting. I kept returning to a magnificent sandstone outcrop bordered on one side by sculpted rock that would have left Frank Lloyd Wright tear-eyed, wondering all the while how I could incorporate that splendid sandstone into my dwelling. Over the next few months my vision of a house on the rocks slowly sharpened. Finally, on a chill November morning, I put my coffee down and started building upon a foundation that was set in place 250 million years earlier.

—

That the sandstone was not level was of little concern to me, for I had seen many an old farmhouse with as much pitch to the floor. Nor was I concerned about the cracks in the rock. I could use the larger fissures to anchor the walls and then employ the natural step along the north edge of the outcrop as a stove hearth. And behind the hearth the artfully sculpted and deeply undercut ledge could jut into the room to become the centerpiece of my house—or at least something to sit on. All I had to do was fashion my walls around what nature had already given me.

So I gathered stone from the land and started work.

One by one I selected rocks, brown on red, red on gray, and scrambled 250 million years of geologic history. I fit them and cemented them, and closed the gaps around the sandstone outcrop. Once I had leveled out the natural footing and anchored planks to the top of the stonework I finished the walls in conventional wood framing—six of them, in an asymmetrical hexagon that fit the slab naturally. I worked alone, with hand tools and native instinct, drawing upon experience I had accumulated from watching other builders and tinkering with my previous homes. It was creative carpentry, to be sure, but out of it emerged an inherent vitality, a soul, expressed in every rock, plank, and pillar.

Slowly the house came together, its walls and gently sloped roof materializing from the remnants of razed buildings, much as new trees sprout from old stumps. I gathered odd construction materials wherever I could, rejecting nothing that might keep weather out or let light in. I watched the classified ads, scoured flea markets, followed demolition crews around. And I found a use for all manner of discards: The framework of a vintage utility trailer braced the corners of my new house; glass from the display counter of an old Navajo trading post made a floor-to-ceiling window; weathered board siding torn off the Arizona Bible Mission graced the walls inside as well as outside. Plywood shipping crates, a used skylight, doors from an old bathhouse: The materials list reads like a collector's guide to junkyard treasures.

Like the ontogeny of a tree, though, the end result bore little resemblance to the seed. The rough boards fit into the native rock and the house began to grow into something coherent, something extraordinarily pleasing, something that struck a chord with everyone who visited. The natural feel of the interior was both inviting and coddling, the wraparound windows protective but not isolating. The woodwork seemed to radiate warmth from other lives in other places, yet the whole appeared as if it had always stood on that rock. People began to take it seriously.

With the raising of the house something else began to emerge. As I pulled up rocks and sifted through old boards, I uncovered a new level of contentment with myself and my place in the world. Living here, I take pleasure in the water I have because I harvest and filter it myself. My roof has become my watershed, and in the scant rainfall of the high desert I find ample supply. I store solar electricity by day, eat dinner by candlelight, and have sufficient power in the evening for my computer, music, and lights. I cook and refrigerate with propane, and nothing in my house hums or whirs. I compost organic waste and grow flowers. I heat my shower with the sun, bathe in a warm sauna, and drain wash water to the outside plants. In winter I burn an armload of wood before the sun is up, and another after the sun goes down.

Year after year I have found comfort and inspiration on that rock, participating fully in the details of living, always aware of what is going on in the world around me. I know, simply from my daily routine, how much rain falls from passing storms, what phase the moon is in, how the constellations change with the seasons, when the cicadas emerge and the claret cups bloom, what day the nighthawks arrive in spring, and when the piñon nuts are ripe in the fall. It is a modern, Thoreauvian existence, and it is entirely satisfying.

Peter Marchand is a field biologist and writer presently studying natural ecosystems at the Catamount Institute on the north slope of Pike's Peak, Colorado. He returns to his home in Arizona whenever he can.

ONE SUMMER MORNING, I woke to find a canyon wren inside my house, perched on a ledge near the stove. Canyon wrens please me to no end with their energetic spirit and unrestrained exuberance for exploration. I have seen them squeeze through the narrowest of rock crevices and disappear into darkness, only to pop out again somewhere nearby with a bold, triumphant chirp. This one had entered my house through a roof overhang, where a board had warped to create a narrow opening. Once it was

between the rafters, the wren found enough space under the insulation to work its way to an unfinished corner of my ceiling. There it dropped into the room and made itself at home.

From plant to rock to windowsills and mantle, it flitted about, probing and exploring as if my house were part of its regular territory. The bird was familiar and comfortable inside, and there was little doubt that it had been here before. I had seen traces of it occasionally in the past — a tiny chestnut

feather on the bureau, a little splash of white on the rock — but until now it had remained more mythical than real. Satisfied after a few minutes of exploration, the wren darted back to its point of entry, scooted under the insulation, and, outside a moment later, dropped past my window with its exultant trademark call. That tacit declaration made it clear that as deeply and inextricably personal as this house is, it is not entirely mine, and never will be.

Above, our main house and offices with the solar hydronic collectors, and bidirectional Internet satellite dish on its roof. Right, our straw bale greenhouse/bathhouse with solar collectors for hot showers and washing. In the front right, some of the motley collection of 72 PV modules that energize Home Power.

Above right: Master bedroom, which opens to the inset, second story, outside deck seen in the photo above.

HOME POWER MAGAZINE HEADQUARTERS GETS A REBUILD

Richard Perez

Our woodstove, which uses a secondary catalytic converter to increase fuel efficiency and reduce pollution. Last year we burned less than ½ cord of wood, thanks to the solar heating systems.

I first saw Home Power *magazine in the '80s. It was a funky looking yet technically loaded and serious journal of (mainly) solar, wind, and water-generated electricity. Not only has it survived, but it's gotten increasingly better. It's now a four-color compendium of the latest in home power. Richard and Karen Perez are the heart and soul of* Home Power, *and after some years of living in funky sheds in the woods, they built their own home-powered home/office/hangout in the Oregon woods. I find it just amazing to look at a place like this, off-the-grid, its heat and power provided by sun and wind (and firewood). And they are running the computers and network that produces their magazine from the same clean electricity. These guys are walkin' the walk!*

Here is Richard's brief history of the magazine and an article on their home:

WE STARTED *Home Power* in 1987 and to date have published 90 issues. Prior do doing *Home Power*, I spent ten years as an installing dealer of PV systems. I solarized our predominately off-grid neighborhood by installing over 200 systems. I realized that folks had no idea of what current solar energy technologies could do for them — they were still running generators to power their off-grid homes and businesses. I also saw an emerging renewable energy industry which had no way to contact their potential customers. Hence *Home Power* was born.

Currently we are entering our 15th year of publishing. Including folks who download our current issue for free from our web site, we have over 100,000 people reading each issue. We print 38,000 copies in our paper edition and about ⅔ of these are sold on newsstands worldwide.

For many years, we lived and worked in a 560-square-foot "plywood palace." This uninsulated building was chock-a-block with the necessities of life and computers. Our site is six miles off-grid, and we've been powering all our electrical stuff using solar and wind electricity for decades now. In summer of 2000, we did a total rebuild — the original cabin disappeared into the center of a new 2,300-square-foot building.

The new building has two stories. The ground floor is split-level, with a four-foot drop along its east/west axis. Thus the building follows the contour of the south-facing hillside on which it rests. The building was designed and constructed by the *Home Power* crew.

Energy efficiency was our major design criteria. We employed both passive and active solar heating techniques. On the passive side, we insulated the hell out of the building — R-30 in the walls and R-60 in the roof. We installed many south-facing, double-glazed windows, a few east-facing windows for an "early morning wake-up," and very few windows on the west and north sides of the house.

Left: A view of the living room from the west. The red tile on the floor covers the solar-heated, concrete slab. Above: The wide open spaces are the reason we live in the mountains. Our nearest full-time neighbor is more than six miles away.

Computer-designed overhangs prevent all these windows from overheating the building during the summer. On the active side, we installed four, 4-by-8-foot solar hot water collectors on the roof. These collectors directly heat a six-inch-thick, concrete, thermal slab on the ground floor. The combination of passive and active solar heating, and super insulation have reduced the amount of wood we burn in our backup heater from five cords per winter to less than one-half cord per winter. We increased the size of our home/office by a factor of four and reduced our wood consumption by a factor of ten, which overall increased performance by forty times.

Besides finally having enough space to not be crowded, the new building is very comfortable — warm in the winter and cool in the summer. We are located at 3,320 feet elevation in the Siskiyou Mountains of southwestern Oregon. It gets cold here in the winter. Nighttime temperatures are often in the teens, and it's not uncommon to have several feet of snow on the ground. Inside the building, it's always cozy. The thermal slab stores enough heat for around four days of continuously cloudy weather. Proof of wintertime performance is that all our dogs and cats prefer to sleep on the solar thermal slab instead of any other place in the house.

During the summer months, when the outside temperature is often in the high 90s, the inside temperature never rises above 76 degrees. We open the many operable windows after sunset and allow the cool mountain air to chill down the house. In the mornings, we simply close the windows and allow the super insulation to keep the house cool during the day.

 Richard Perez, *Home Power*, P.O. Box 520, Ashland, OR 97520, 541-941-9716
richard.perez@homepower.com
www.homepower.com

Another smaller cabin that houses one of our friends

View from the sunken living room up into the dining area

Kitchen with just about every appliance Karen needs to do her gourmet cooking

Dining area with a table that will seat ten people

Power room, which houses our batteries, inverters, and other renewable energy equipment

One of the three offices at our home. In all, these offices house five of the computers we use to make Home Power *magazine.*

CABAÑA EN ESPAÑA

Dear Lloyd,

Here are photos of my cabin. I was inspired and helped a lot by the books *Shelter* and *Shelter, Shacks and Shanties*. *Shelter* is a marvel, I like the spirit of the '60s and '70s, which is reflected in it.

It has always been my dream, since I was a child, to build a cabin of wood with my own hands, *mi casa*. Finally I was able to buy some land in Miraflores, a town in the mountains near Madrid, and build a small cabin. Now I've finished construction. The only thing lacking now is finishing parts of the interior. . . .

During construction I felt like a naughty child, or like an adolescent revolting against a mechanized and absurd world, seeking a life in the country in a small handmade cabin, with few fears and great hopes, even though simple. I planted a small garden, I sit and listen to the sounds of the creek, or play the violin inside my little house — nothing more.

Best wishes,
Enrique Sancho Aznal
Madrid, Spain
(Translated from the Spanish)

" . . . I felt like . . . an adolescent revolting against a mechanized and absurd world, seeking a life in the country in a small handmade cabin. . . ."

CABIN IN TENNESSEE

We print our book Shelter *at the Quebecor Press in Kingsport, Tennessee. On books like this, with a lot of photos, I always go to the printers to watch the first run. One of the press men was Garry Crawford and when he saw the photos in* Shelter, *he told me about a little cabin he had built out in the woods, a "childhood dream." (Guys just wanna have fun!) Here is his story of the cabin, built in Hawkins County, Tennessee, by Garry, his brother Larry, and their buddy Leonard Lamb.*

Here's how our cabin was built:

- Builders — two brothers, no experience; 1 ex-merchant marine, no experience
- Time — one year (weekends)
- Expenses — $180.00 for two acres in the woods; approximately $150.00 for materials: nails, mortar mix for chinking. All lumber and metal for the roof was scrap material found all over two counties.
- One chainsaw borrowed from a good friend

The cabin is warmed by a wood-burning cookstove. Water is from four springs. Lights are kerosene. Windows are from an abandoned chicken-house. The front and rear eaves of the cabin are built with wormy chestnut lumber from a collapsed barn.

This has been one of the most enjoyable projects I have ever undertaken. Not only has it fulfilled three grown mens' childhood dreams, but it has given me confidence to start a new addition — coming soon.

– Garry Crawford

57

JOANNE'S HOUSE

Joanne Kyger is my neighbor, a poet, and an elegant lady. Her house, an old cottage she bought in 1970, reflects her travels to various parts of the world and has a wonderful feeling inside. Everywhere you look are things of beauty: a Tibetan tanka, a Balinese painted calendar, lots of paintings, dozens of baskets, healthy green plants, Japanese vases and laquered plates. There's a mirror from Guatemala, the smell of incense, and a bookshelf with hundreds of books. The old water-stained shingles on the roof show through in the living room, and there's a woodstove for heat.

To enter the property you walk through a tunnel in a massive 60-year-old cypress hedge sculpted by Joanne's partner Donald Guravich. In the garden there are places to sit and watch families of quail scurry through, and to look at the different plants and bushes and trees that are all carefully tended. There are also multiple varieties of apples growing, which Donald has grafted onto old trees, and they ripen from August until October.

In a recent magazine article, Joanne was called a "poet's poet," and Penguin has just published her most recent book, a collection titled *As Ever*.

FRIDAY NIGHT

In pale blue dusk   sky   Moon
is nice light gold.   Oh where
are you going
my favorite friends in a flock   Gold crown
song is going north
for the summer   has different
seeds up there   up there friend moon
is getting larger.

April 26, 1991

Photos by Janet Holden Ramos

Renée and daughter Felícidad

THE HOUSE THAT RENÉE BUILT

Renée Doe built this house in a Northern California valley in the early '70s. She and five other families had bought 50 acres on which to build homes and grow vegetables. For building materials they bought two old buildings from the county for $25 each, as well as a flatbed truck; they tore down the buildings and ended up with a bunch of good quality redwood sheathing, as well as oak flooring. Architect Steve Matson designed the house to Renée's specifications: "I wanted a seven-gables-looking house with steep roof angles, and I told Steve I wanted the pieces of lumber to be small enough for women to lift."

Renée lived in a tent with two of her three kids while she and friend Maggie Cooley worked on the house. "We figured, if you can sew, you can build a house." They started building in July and Renée and kids moved into the house by November, using a wood cookstove for heat and cooking. Windows were salvaged from a dumpster in nearby San Francisco. "We carried water up to the house in buckets, and we dug an outhouse." The kids bathed in the creek. "It was awful; by fall it was freezing cold." I remember seeing the house under construction and thinking how complex it looked, but when it was finished, it all worked out. The kids survived and made it out into the world and Renée and her partner Brent Anderson now live in what turned out to be a cozy home with good vibes.

COLOR IN THE CARRIBEAN
Renate Bonn

Alameda Naval Base

El Cerrito

Berkeley

El Cerrito

Berkeley

San Francisco

Sausalito

San Francisco

SAN FRANCISCO
BAY AREA COLOR

San Francisco

Oakland

Berkeley

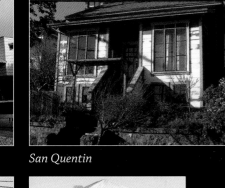

Berkeley   *San Francisco*   *San Francisco*   *Berkeley*

Berkeley   *San Francisco*   *San Quentin*

San Francisco   *San Francisco*   *San Quentin*

Berkeley   *San Francisco*   *Richmond*

San Francisco   *San Francisco*   *Oakland*   *San Francisco*

"It ain't what you eat,
    it's the way how you chew it."
                —Little Richard

Janet Baer in her Sonoma County kitchen

CALIFORNIA KITCHENS

Sam and Nidia Birenbaum's kitchen in their
beachfront house in Malibu

SHED ROOF

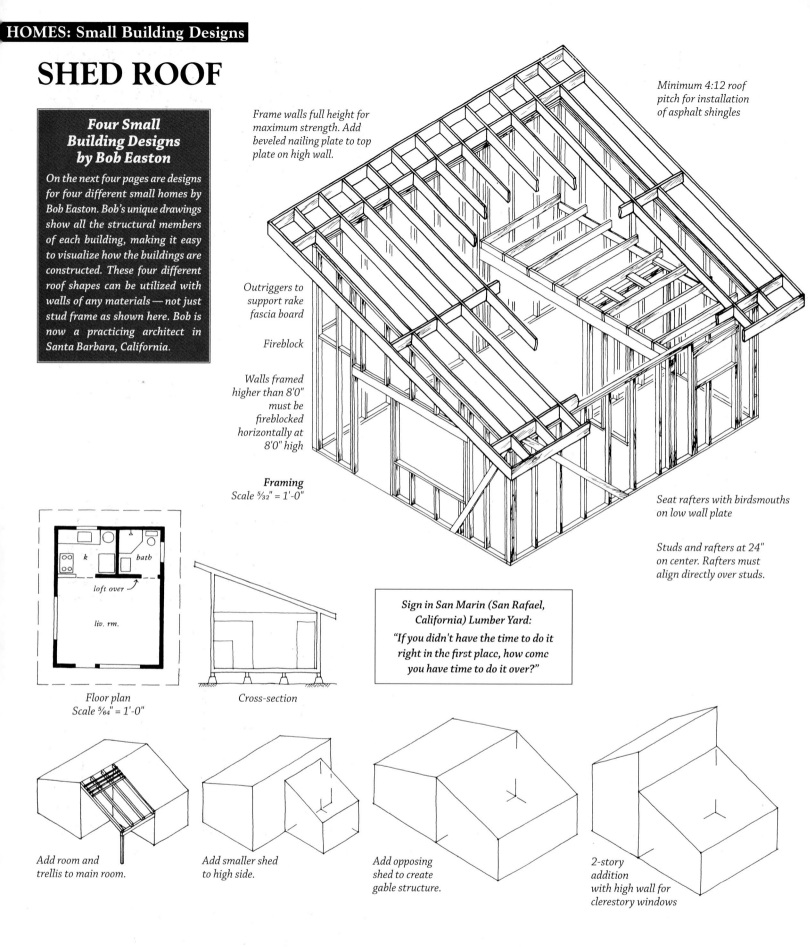

Frame walls full height for maximum strength. Add beveled nailing plate to top plate on high wall.

Minimum 4:12 roof pitch for installation of asphalt shingles

Outriggers to support rake fascia board

Fireblock

Walls framed higher than 8'0" must be fireblocked horizontally at 8'0" high

Framing
Scale 5⁄32" = 1'-0"

Seat rafters with birdsmouths on low wall plate

Studs and rafters at 24" on center. Rafters must align directly over studs.

Sign in San Marin (San Rafael, California) Lumber Yard:

"If you didn't have the time to do it right in the first place, how come you have time to do it over?"

k

bath

loft over

liv. rm.

Floor plan
Scale 5⁄64" = 1'-0"

Cross-section

Add room and trellis to main room.

Add smaller shed to high side.

Add opposing shed to create gable structure.

2-story addition with high wall for clerestory windows

A SHED ROOF is a simple shape to build, sheds water and snow better than a flat roof, and is a good shape for later additions or extensions. It is also a good shape for adding to an existing building. Clerestory (high) windows are often installed on the high side of a shed roof building, allowing high light to enter without the waterproofing problems of skylights. *(See drawing above right.)*

Shown above is a small shed building with a six-foot-wide loft. The smaller drawings show additions to the shed shape. When building overhang on shed, nail rafters securely to top plate with 4-16d nails.

HIGH GABLE

Two basic ways to frame a gable roof:
1. With ridge beam as shown here, which allows for open ceiling, and with ceiling joists dropped below plate level (also shown here). Ridge beam should be sized by engineer or checked by building inspector. With ridge beam system, end walls must be rigidly braced — plywood is best. Carpentry must be accurate, joints tight, nailing adequate.

2. With ridge beam and ceiling joists at plate level. With loft floor, ceiling joists must be sized as floor joists.

12: 12 roof pitch

Studs 16" on center; rafters 24" on center, cross ties 4'0" on center

Frame plate level all round — locate fireplace at end wall or near center to avoid tall unsupported chimney.

Framing
Scale ⁵⁄₃₂" = 1'-0"

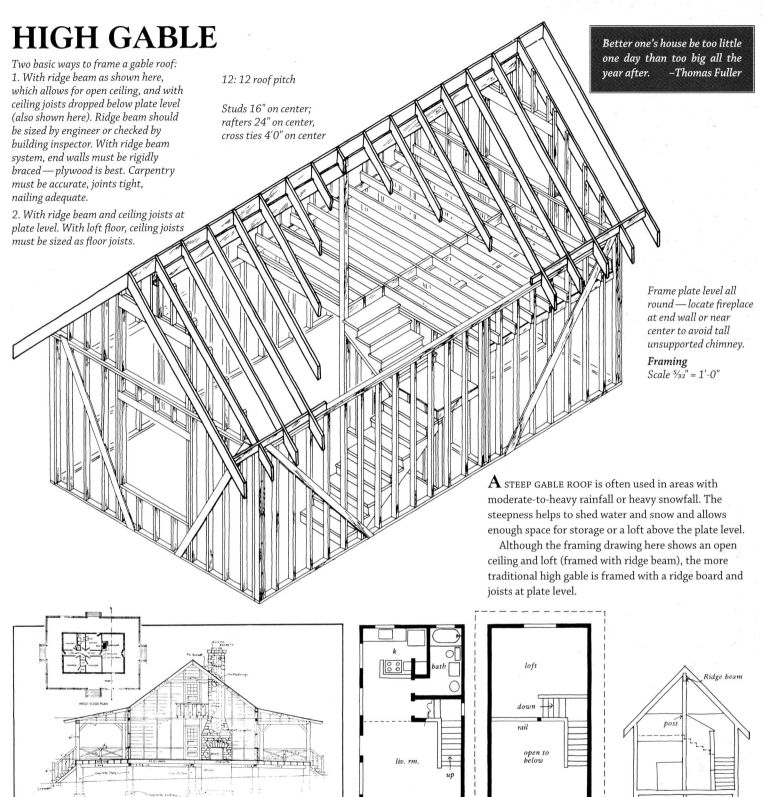

A STEEP GABLE ROOF is often used in areas with moderate-to-heavy rainfall or heavy snowfall. The steepness helps to shed water and snow and allows enough space for storage or a loft above the plate level.

Although the framing drawing here shows an open ceiling and loft (framed with ridge beam), the more traditional high gable is framed with a ridge board and joists at plate level.

Plan and section of bungalow in Pennsylvania

Main floor plan
Scale ⁵⁄₆₄" = 1'-0"

Loft plan

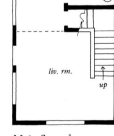

Cross-section

more...

SALTBOX

THE SALTBOX SHAPE is generally associated with the New England states and severe winters. Saltbox structures are often oriented with the high side to the south, the low side to the north. This allows winter sun to hit the high side, and snow (a good insulator) to accumulate on the lower, shallower roof to the north. Snow or bales of hay are often banked against the north side in winter for insulation.

9:12 roof pitch

Studs at 16" on center; rafters at 24" on center

Note: In snow country, check with engineer or building inspector.

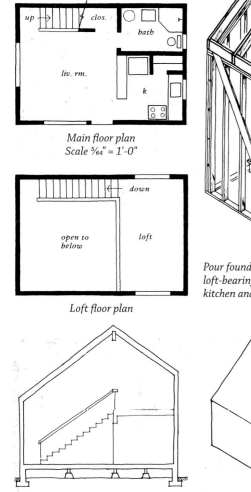

Main floor plan
Scale 5/64" = 1'-0"

Loft floor plan

Cross-section

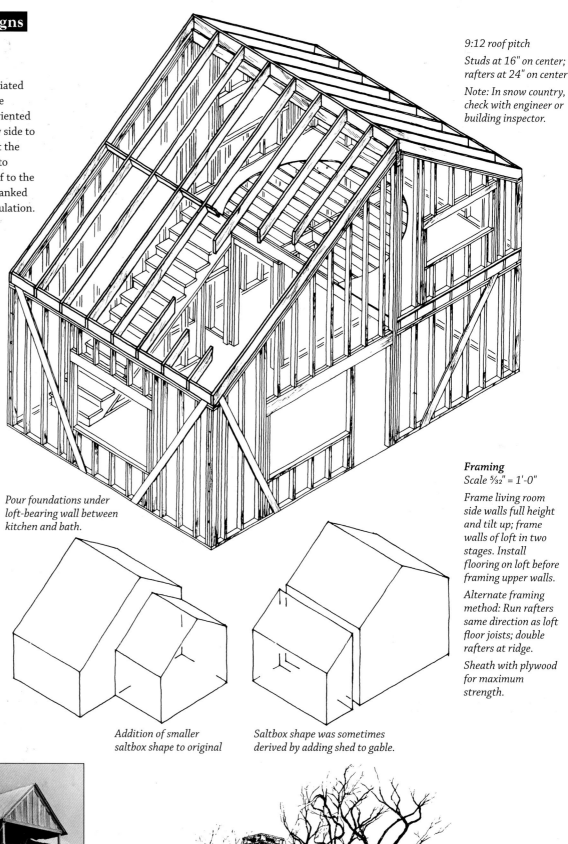

Pour foundations under loft-bearing wall between kitchen and bath.

Addition of smaller saltbox shape to original

Saltbox shape was sometimes derived by adding shed to gable.

Framing
Scale 5/32" = 1'-0"

Frame living room side walls full height and tilt up; frame walls of loft in two stages. Install flooring on loft before framing upper walls.

Alternate framing method: Run rafters same direction as loft floor joists; double rafters at ridge.

Sheath with plywood for maximum strength.

GAMBREL

Gᴀᴍʙʀᴇʟ ʀᴏᴏꜰꜱ are most often found in the eastern part of the United States and Canada. The word derives from the hock (bent part) of a horse's leg, also called a gambrel. The lower part of the roof is a steep slope, the upper part shallower. The break in roof line allows head room in the loft space, and is useful in barns for hay storage (see page 212 for gambrel barn plans), as well as in homes for rooms above plate level.

Minimum 4:12 pitch for asphalt shingles

Cross ties at 4'0" on center

Frame main floor walls, install loft floor joists and flooring, then frame loft walls and roof.

Studs and rafters at 16" on center

Framing
Scale ⁵⁄₃₂" = 1'-0"

> *The best way to realize the pleasure of feeling rich is to live in a smaller house than your means would enable you to have.*
> *–Edward Clarke*

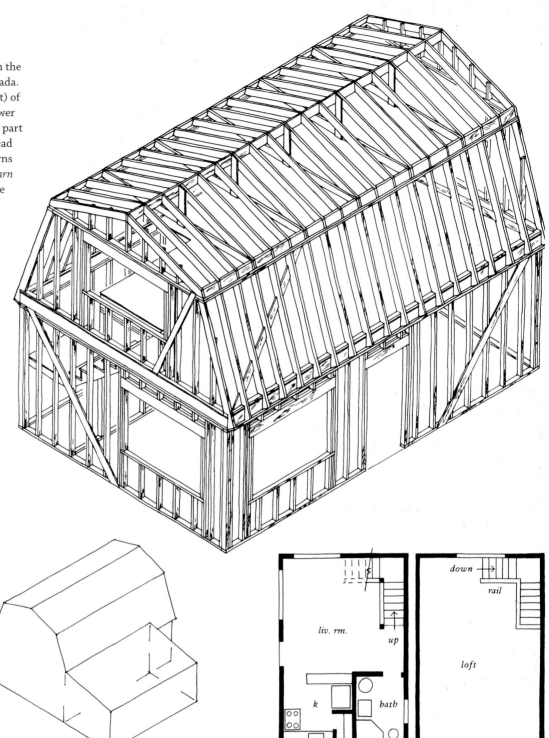

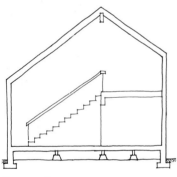

Cross-section

Addition of shed off side

Main floor plan
Scale ⁵⁄₆₄" = 1'-0"

liv. rm.

up

k

bath

down

rail

loft

Loft plan

TINY HOUSES

by Lester Walker

The six little house plans shown here are from Tiny, Tiny Houses *by Lester Walker. Lester is a rarity —an architect who not only has designed these little houses, but has drawn clear and useful plans that he shares with others. There are 40 designs in this unique book.*

THE FIRST TINY HOUSE I remember seeing and categorizing as a tiny, tiny house was a complete surprise. In the summer of 1963, I discovered one while hiking along what seemed to me to be a very treacherous, untraveled animal trail on a remote part of Maine's coastline, about an hour east of Cutler. I couldn't imagine how anyone might have transported materials to this spot without having lugged them over windswept cliffs and slippery rocks. But there it was, a tiny little gable-roofed cabin no larger than 8' × 10' built entirely of tarpaper and driftwood, complete with an Adirondack-style built-in twig bed, a perfect little kitchen that used water from a nearby spring, and a writing desk under a window facing the sea. Set back about one hundred feet from the ocean on a rocky beach in a small cove, the house was surrounded by cliffs topped with huge hemlock and pine trees. Later, when I got back to town and learned that it was built by a little lady in her eighties who loved nature and solitude, I realized that the art of building was not necessarily reserved for architects and builders. All that was needed, it seemed, was the will. Two years ago, I hiked back to this site with my camera, notepad, and the hope that I could find this little building to include in *Tiny Houses*. No luck. A big storm had apparently blown it away. But this home will remain in my mind as one of the most beautiful buildings I've ever seen. It may well have been the inspiration for this book.

—Lester Walker

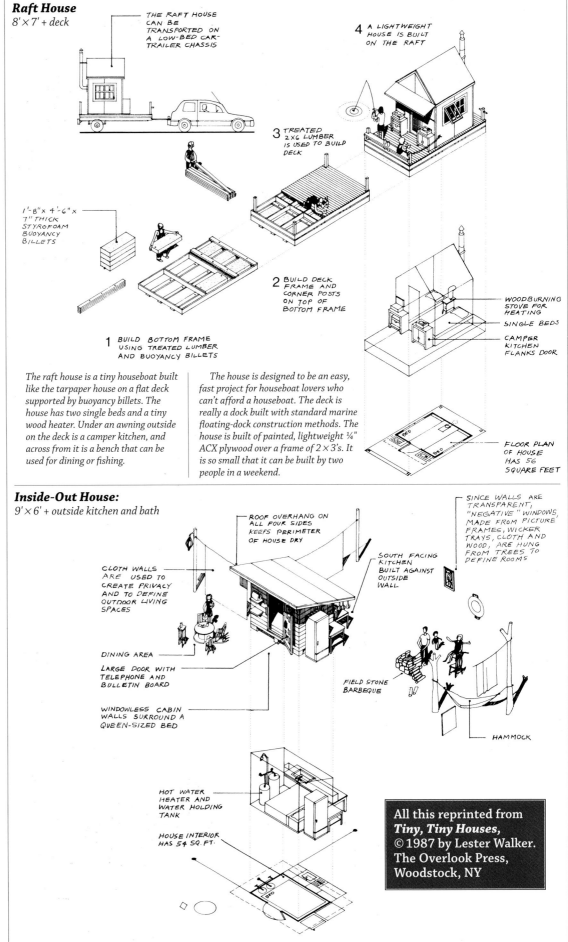

Raft House
8' × 7' + deck

THE RAFT HOUSE CAN BE TRANSPORTED ON A LOW-BED CAR-TRAILER CHASSIS

1'-8" x 4'-6" x 7" THICK STYROFOAM BUOYANCY BILLETS

1 BUILD BOTTOM FRAME USING TREATED LUMBER AND BUOYANCY BILLETS

2 BUILD DECK FRAME AND CORNER POSTS ON TOP OF BOTTOM FRAME

3 TREATED 2X6 LUMBER IS USED TO BUILD DECK

4 A LIGHTWEIGHT HOUSE IS BUILT ON THE RAFT

WOODBURNING STOVE FOR HEATING

SINGLE BEDS

CAMPER KITCHEN FLANKS DOOR

FLOOR PLAN OF HOUSE HAS 56 SQUARE FEET

The raft house is a tiny houseboat built like the tarpaper house on a flat deck supported by buoyancy billets. The house has two single beds and a tiny wood heater. Under an awning outside on the deck is a camper kitchen, and across from it is a bench that can be used for dining or fishing.

The house is designed to be an easy, fast project for houseboat lovers who can't afford a houseboat. The deck is really a dock built with standard marine floating-dock construction methods. The house is built of painted, lightweight ¼" ACX plywood over a frame of 2 × 3's. It is so small that it can be built by two people in a weekend.

Inside-Out House:
9' × 6' + outside kitchen and bath

ROOF OVERHANG ON ALL FOUR SIDES KEEPS PERIMETER OF HOUSE DRY

CLOTH WALLS ARE USED TO CREATE PRIVACY AND TO DEFINE OUTDOOR LIVING SPACES

DINING AREA

LARGE DOOR WITH TELEPHONE AND BULLETIN BOARD

WINDOWLESS CABIN WALLS SURROUND A QUEEN-SIZED BED

SOUTH FACING KITCHEN BUILT AGAINST OUTSIDE WALL

SINCE WALLS ARE TRANSPARENT, "NEGATIVE" WINDOWS, MADE FROM PICTURE FRAMES, WICKER TRAYS, CLOTH AND WOOD, ARE HUNG FROM TREES TO DEFINE ROOMS

FIELD STONE BARBEQUE

HAMMOCK

HOT WATER HEATER AND WATER HOLDING TANK

HOUSE INTERIOR HAS 54 SQ. FT.

One of the most clever houses in this book is this tiny windowless building, just as big as a double bed, built in 1967 by a young couple in Sharon, Connecticut, for shelter while they built the log cabin of their dreams nearby. It's called the Inside-Out House because all the living functions, except sleeping, occur on the outside periphery of the building.

A large overhanging roof protects L-shaped kitchen cabinets on two exterior walls and a shower on a third wall. A big door, usually open for ventilation, occupies the fourth wall. An enormous dining room, as large as all outdoors, is located adjacent to the big door. The living room, equally as large, is located adjacent to the kitchen. These rooms are defined primarily by trees but also by "negative windows" such as wicker trays, picture frames, and pieces of cloth hung from the trees. As David Bain, the owner/builder, expresses it, "Since we could see through our walls, we didn't need to see through our windows."

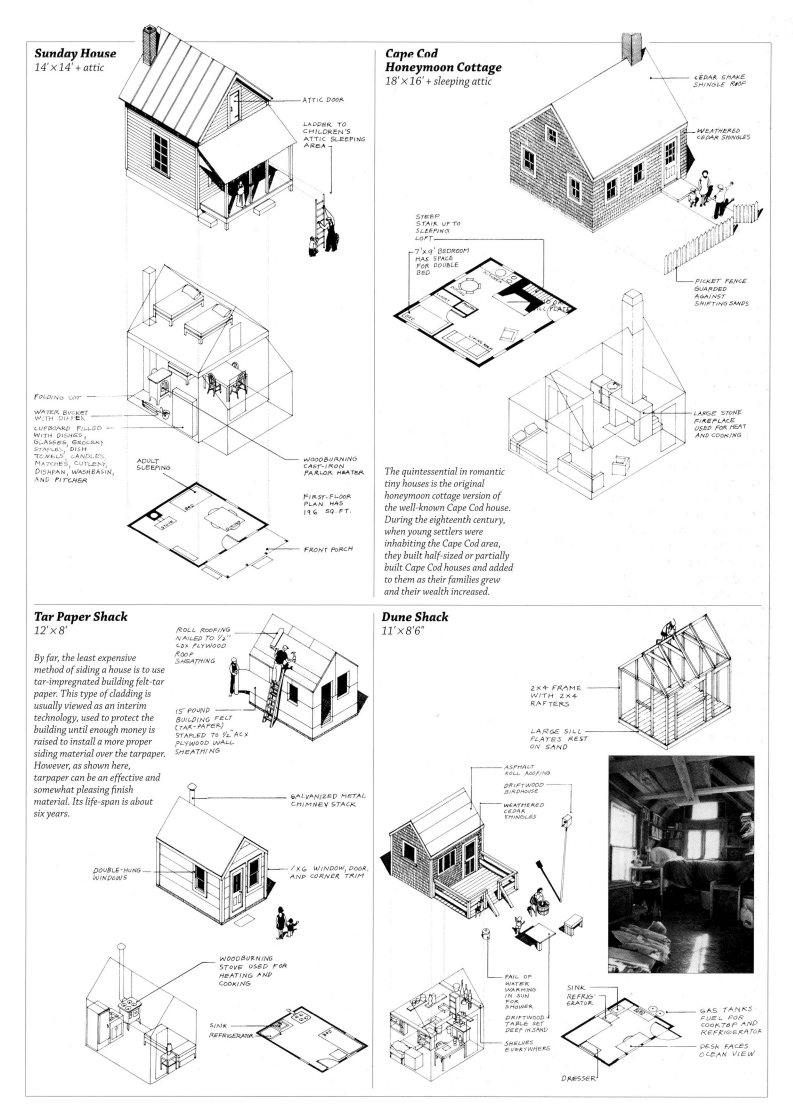

Sunday House
14′×14′ + attic

ATTIC DOOR

LADDER TO CHILDREN'S ATTIC SLEEPING AREA

FOLDING COT

WATER BUCKET WITH DIPPER

CUPBOARD FILLED WITH DISHES, GLASSES, GROCERY STAPLES, DISH TOWELS, CANDLES, MATCHES, CUTLERY, DISHPAN, WASHBASIN, AND PITCHER

ADULT SLEEPING

WOODBURNING CAST-IRON PARLOR HEATER

FIRST-FLOOR PLAN HAS 196 SQ.FT.

FRONT PORCH

Cape Cod Honeymoon Cottage
18′×16′ + sleeping attic

CEDAR SHAKE SHINGLE ROOF

WEATHERED CEDAR SHINGLES

STEEP STAIR UP TO SLEEPING LOFT

7′×9′ BEDROOM HAS SPACE FOR DOUBLE BED

PICKET FENCE GUARDED AGAINST SHIFTING SANDS

LARGE STONE FIREPLACE USED FOR HEAT AND COOKING

The quintessential in romantic tiny houses is the original honeymoon cottage version of the well-known Cape Cod house. During the eighteenth century, when young settlers were inhabiting the Cape Cod area, they built half-sized or partially built Cape Cod houses and added to them as their families grew and their wealth increased.

Tar Paper Shack
12′×8′

By far, the least expensive method of siding a house is to use tar-impregnated building felt-tar paper. This type of cladding is usually viewed as an interim technology, used to protect the building until enough money is raised to install a more proper siding material over the tarpaper. However, as shown here, tarpaper can be an effective and somewhat pleasing finish material. Its life-span is about six years.

ROLL ROOFING NAILED TO ½″ CDX PLYWOOD ROOF SHEATHING

15 POUND BUILDING FELT (TAR-PAPER) STAPLED TO ½″ ACX PLYWOOD WALL SHEATHING

GALVANIZED METAL CHIMNEY STACK

DOUBLE-HUNG WINDOWS

1×6 WINDOW, DOOR, AND CORNER TRIM

WOODBURNING STOVE USED FOR HEATING AND COOKING

SINK REFRIGERATOR

Dune Shack
11′×8′6″

2×4 FRAME WITH 2×4 RAFTERS

LARGE SILL PLATES REST ON SAND

ASPHALT ROLL ROOFING

DRIFTWOOD BIRDHOUSE

WEATHERED CEDAR SHINGLES

PAIL OF WATER WARMING IN SUN FOR SHOWER

DRIFTWOOD TABLE SET DEEP IN SAND

SHELVES EVERYWHERE

SINK REFRIGERATOR

GAS TANKS FUEL FOR COOKTOP AND REFRIGERATOR

DESK FACES OCEAN VIEW

DRESSER

BOBOLINK *Better Shacks and Bivouacs*

ONE WARM SUMMER AFTERNOON I went out with Bill Castle *(see pp. 16–21)* to meet his friend Bobolink. Bobolink had bought his piece of land (in northwest New York state) the year before for $1000. "I gave the lady $300 and paid off the other $700 the rest of the year."

He showed us around the unfinished house, which was cozy and comfortable and then we sat around drinking beer.

"I figured I'd just build a little shack, a place in the country to come back to after travelin' around . . . ."

"What about the building codes?"
"When I started there was no uniform code here, so they never gave me any grief."

"Did you draw any plans?"
"I just did these sketches, not to scale, of different parts of the room, then taped them together and showed them to my girlfriend."

"She doesn't hassle you to get it finished?"
"No. In fact we just tied the knot last Thursday. We're a perfect team."

"Well, it looks like you got the hard part done."
"No, man, the hard part is doin' anything more. I sit down, kick back and . . . well, it seems too hot today . . . how about a beer?" *(laughter)*

72

NATURAL MATERIALS

IN THE EARLY '70s, after building geodesic domes and experimenting with plastic building materials, I came to the conclusion that the less molecular rearranging a particular building material has, the better it feels to be around. The key word is *feels*. Wood, adobe, straw, earth, stone, bamboo—these materials feel good.

There's been a revolution in the use of "natural building materials" in the last 30 years. Builders are choosing materials for sustainability, for less drainage on the earth's resources, for local availability. A number of people have told us that *Shelter,* published in 1973, with its photo of a straw bale barn in Nebraska, and pages on wood, stone, adobe, thatch, and bamboo, had a lot to do with sparking interest in these materials.

By now a lot of construction techniques have been worked out and there is a large network of builders out there, using (and communicating about) natural materials. On the following 23 pages are some examples of people building this way. *Buen trabajo, amigos!*

Ongoing and never-ending remodel of early 1900s adobe ranch/farm house

Kitchen fireplace, handcrafted from molded clay, straw and pumice

Kitchen sitting area, corner seat of adobe, walls painted with homemade casein paints

BILL & ATHENA STEEN AND THEIR HOUSES OF MUD & STRAW

On a hot day in late July, 2002, I drove south from Tucson, heading up into the high desert to visit Bill and Athena Steen. Bill and Athena, authors of the *The Straw Bale House* book, a best-seller and precursor of the straw bale building movement, had done an impressive mud/straw/bamboo series of buildings with villagers in Ciudad Obregón, Morelos, Mexico, and I wanted to do a story on it for *Home Work*.

Another reason for the visit was the chance to meet photographer extraordinaire Yoshio Komatsu, author of the stunning book *Living on Earth*, who with his wife Eiko was visiting the Steens at that time.

The Steens live on a 40-acre homestead 70 miles southeast of Tucson (15 miles by crow-flight from Mexico) and at the end of a dirt road. They bought the land in 1985 and Bill converted a run-down shack into what is now a gracious and comfortable hacienda, with adobe walls and floors of Mexican tile. These days Bill and Athena use their homestead to host a series of workshops on straw bale building, natural wall finishes (main ingredient mud), earthen floors, clay ovens, and harvesting and cooking agave and prickly pear.

What I expected was to work with the Steens on their Mexico project, What I didn't expect was such an elegant house, set alongside a creek, in a place with Feng Shui up the kazoo, with good vibes, sights, colors, smells — the essence of wonderful shelter — plus there was a series of experimental earth buildings, each one a delight, and with a variety of textures, colors, and construction innovations.

Athena, Yoshio, and Bill

Shade screens made by laying water reed mats over prefab livestock panels used for fencing

View of main guest and bed/breakfast building, looking across Turkey Creek

Clay baking ovens; one on left is hand-molded clay and straw, Quebec style; one on the right is in traditional southwestern "horno style" and made of adobes.

Bill, Athena, and their three kids — Benito, 11; Oso, 10; and Kalin (Bug), 2 — are way out there in the desert. The older boys are home-schooled. Bill and Athena work on their building techniques, writing, photography, and teaching. Bug happily wanders around all day, barefoot and bare-assed, whacking a golf ball with a driving iron and amusing himself in amusing ways. One day he came up to me with a salad bowl on his head, a straight face, and watched for my reaction.

I slept in an adobe-walled bedroom, with two screened doors opening out into a bamboo grove in the garden. The first morning I hiked up on the hill to watch the sunrise, then came down and shot pictures. The second night there a storm hit, and thunder, lightning, and the good smells of the desert came in through the screen doors next to my bed.

Tiles by the Mexican artist Davalos, bought on one of their tequila, blanket, papaya, and taco runs to Nogales

Multi-functional corner bed in living room. Remove mattress and it becomes a desk. Bug (the two-year-old) has a playroom complete with window underneath and little step often gets used as seat.

WEB: www.caneloproject.com
EMAIL: absteen@dakotacom.net

Fish pond from galvanized metal stock tank, effortless and beautiful

Bug with salad bowl hat

more…

Paint and plaster studio, also called the rain chain building, almost finished. Straw bale walls with clay/straw plaster on exterior

Belled-out corner similar to detail used in mud construction in Africa

Arched window in straw bale walls, clay-plastered wall finished with red wax and pigment. Interior of window is white clay plaster with micaceous sand.

Sponge being used to finish freshly applied plaster

Gallo wine bottles set into straw bale walls with clay plaster

Guest cottage, one of our favorites. Like everything else, still under construction. A variety of clay plasters. Exterior has straw/clay plaster in front and lavender clay finish on wall around window. Bamboo, clay and straw shelves to rear were molded into place. Interior has polished clay walls.

Clay and straw molding around window in straw bale walls. Arch over the window was formed by making a lightweight frame out of split bamboo and covered with mixture of clay and straw.

Structure ("bus stop") used for seating during workshops. Local juniper poles, straw bale walls, seat is from blocks made from sawdust and clay during a workshop. As soon as they came out of mold, they were placed into seat wet to be able to form curves. Behind it is straw bale storage shed for lime and clay putties so they don't freeze; other dry plaster materials are stored on the outside.

Interior of guest cottage, by-product of first straw bale workshop/happening in 1990. Adobe wall/seat divides bathroom space from living area. Back side of the wall forms lime-plastered shower.

Future bath house being used as temporary storage for tools and stuff that have no other place to go

Above: Bike shed, with clay/straw plaster on exterior, lime plaster on inside and ceiling of petates, or mats of carrizo and cattails

Cut-in-half clay pot makes perfect indirect light fixture.

Other "bus stop," used for workshop seating. Both bus stops are invaluable places for interns and the like to refine their skills and experiment without us having to pull our hair out. The souls of many a good intern are embedded in multiple layers of paint and plaster.

Mixing our basic clay/straw plaster that we evolved while working in Mexico. Screened local clay soil is sprinkled into water until water is covered, then allowed to sit for a few minutes so clay soaks up water. Messing with it sooner makes for lots of clumps. We then add enough chopped straw to make a mix that has body and can be applied thickly on walls without cracking. Our exterior plasters are applied up to 2 inches thick in one single application.

Bottom right two photos show "harling," where the lime plaster slurry is thrown on the wall using a "harling" trowel. Great durable and weather-resistant finish

Bill with intern Lori Wright standing on arch made of clay and straw

more...

Patio of Save the Children Office Building, palm-thatched porch on the interior; the floor is a pattern carved by spoon into fresh concrete.

CASAS QUE CANTAN

Reception room of Save the Children Office Building, frescoed lime plaster on walls. Blue color comes from azul anil, *a blue pigment commonly found in the* Dulcerias *or candy stores.*

WHEN WE FINISHED our first book, *The Straw Bale House*, we had used up our available credit and knew that we wouldn't see any royalties or the like for some time. Things looked a little grim, but we were ready for anything that looked interesting and exciting. The first offer that came our way was from Save the Children Foundation in Ciudad Obregón, Mexico—a modern agricultural town in the southern part of the state of Sonora. We took the offer —a place to stay, gas, tacos, and any repairs our aging Suburban needed to get us there

and back. What ensued was an eight-year love affair with the most unlikely of places and a big extended family with whom we have formed a deep and lasting friendship/ partnership.

We joined together in what became an ongoing exploration of every type of local building material imaginable, mostly local, natural, and recycled. We combined and re-combined them into a series of experimental small homes and an office building for Save the Children. They have all come to be referred to as "Casas que Cantan," or

"Houses that Sing," after the exquisite book by Mexican photographer *La Casa que Cantá*. More than anything else, the work was fun, lots of it. People often get the mistaken idea that we went there to help poor people. It would be more correct to say that we were the ones who benefited the most, for the emptiness of our modernized poverty got filled in in countless and unimaginable ways by people who were in many ways richer than us.

–Bill and Athena Steen

Athena with her plastering crew — part-time daughters Elizabeth, Juanita, and Maria Elizabeth

WEB: www.caneloproject.com/cqc.html
EMAIL: absteen@dakotacom.net

Library of Save the Children Office Building. Vault formed by tensioning carrizo reeds and covering them with an insulating mix of straw and clay, finished with a concrete shell. Book shelves made by the women and kids out of molded straw and clay. Walls are finished with beautiful red clay from nearby colonial town named Alamos.

Arched entryway to Save the Children Office Building, built using blocks of clay and straw

The Save the Children Office Building, stacking bales . . .

Emiliano Lopez unloading bales by moonlight

. . . mud applied, south side . . .

. . . finished north side

Entrance, sunflower

Truth window (straw to be cut away)

Interior arches

One-room experimental house of straw bales, built for a family using donations put together by folks all over world

Clay/straw plaster being applied over straw bale walls of Save the Children office

Straw bale herbal apothecary, Sue Mullen, Gila, New Mexico

Straw bale chicken house, Steve MacDonald, Gila, New Mexico

Penny Livingstone and James Stark in front of their straw bale vault in Northern California

Straw bale chapel on a Sedona, Arizona ranch, built in mid-90s

Earthbag builder Kaki Hunter, Moab, Utah

NATURAL BUILDING

Photos by Bill Steen

BILL STEEN started shooting photos in the late '60s when he was in college, influenced by photographer Minor White. He continued in the early '80s, shooting mostly landscapes; in the mid-'80s, when he and his wife Athena started building with straw bales, he started documenting their work "out of necessity." At times it got a little confusing: "One hand would be covered in mud, and I'd have a camera in the other." Two books featuring his photos have been published: the best-selling *The Straw Bale House* and *The Beauty of Straw Bale Houses*.

Above and below: Straw bale studio, Lane McClelland and Lauri Roberts, Descanso, California

Brick dome in Hesperia, California, by Nader Khalili, Author of Ceramic Houses

Carol Anthony and crew, straw bale workshop, Tlaxcala, Mexico

Replica of mound builders' homes, Evansville, Indiana

NATURAL BUILDINGS
Photographs by Catherine Wanek

SINCE DISCOVERING straw bale construction in 1992, Catherine Wanek has traveled widely to spread the straw bale gospel, and documenting traditional and modern examples of natural building. She co-edited *The Art of Natural Building* in 2002 and wrote and photographed *The New Strawbale Home*, published in 2003 by Gibbs Smith, Publisher. Catherine and her husband Pete Fust live in Kingston, NM, where they manage the Natural Building Colloquium – Southwest, and run the historic Black Range Lodge *(blackrange@zianet.com)*.

In Brittany, France, owner-builder Elsa LeGuern designed a straw bale home for herself with wide overhangs to protect the bales from storms blowing in from the Atlantic Ocean. The framework is a rectangle, with curved straw bale walls.

Welsh furniture maker David Hughes built this charming thatched timber-frame workshop, choosing the organic shapes of oak trees that wouldn't suit more rectilinear structures.

At the Lama Foundation, a spiritual community near Taos, NM, a forest fire destroyed most of the existing structures in 1996. In 1999, an event called Build Here Now was organized to help their reconstruction efforts. This passive solar straw bale residence has interior adobe and straw/clay walls for thermal mass, and was finished with earthen floors and plasters. The timber-frame structure, now known as, "The Treehouse," was designed by Sun-Ray Kelly, and utilizes ponderosa pine trees killed in the fire.

Thierry Dronet built this fairy-tale hybrid of straw bales and cordwood masonry, topped with a "living roof," as his workshop and stable for two horses in eastern France. Bale walls act to retain the hillside, with a plastic sheet barrier and a "French drain" to wick away moisture. Time will tell whether this practice is advised.

Lars Keller and a vaulted straw bale dormitory at an agricultural college in Jutland, Denmark. Exterior is lime plaster.

The "honey house" by builders Kaki Hunter and Doni Kiffmeyer in Moab, Utah. This dome/vaulted structure was constructed from earth-filled sandbags and plastered with earth and lime plasters.

Chickens at the Black Range Lodge in Kingston, NM enjoy sculpted cob nests in their straw bale chicken coop.

MUD DANCING
Cob Construction with Ianto Evans and Linda Smiley

Iᴀɴᴛᴏ Eᴠᴀɴꜱ ᴀɴᴅ Lɪɴᴅᴀ Sᴍɪʟᴇʏ are the foremost proponents of cob building in North America. They run the North American School of Building (headquartered in Coquille, Oregon), which provides workshops and apprenticeships in cob and natural building and the Cob Cottage Company, which publishes books on the same subjects. Ianto and Linda live in the cob complex shown here, which forms a semi-circle around a large, beautiful, very productive vegetable garden that looks down on a lake in the Oregon woods.

The Hand-Sculpted House, by Ianto, Linda, and Michael G. Smith, has recently been published by Chelsea Green, and is the definitive work on building cob cottages. Construction of Heart House, as they call their home, is covered thoroughly in the book. We won't go into the building details here — it's covered in their book — but briefly, the foundation is an 18″ rubble trench with a 2″ drainpipe, followed by a knee-high rock wall that was dry-stacked, then mortared to keep out rodents. The cob was placed on top of that. It was (and is typically) mixed by laying a tarp on the ground, and using clay-rich soil, coarse river sand, long-stemmed wheat straw, and water. The mixture was then stomped upon with bare feet, often with music — drums and flutes — in a ritual they call "mud dancing."

It's a charming house, built, as they say, ". . . by taking the ground from under your feet and turning it into a building."

www.deatech.cobcottage
Cob Cottage Company, P.O. Box 123, Cottage Grove, OR 97424
541-942-2005

Fires are built in the drum at left; heat flows through the cob window seat and up chimney in wall at right.

Ianto and Linda's complex, with the Heart House at the center

Looking down through garden to lake

Ianto in the morning sun

Beautiful garden!

The day I visited, Linda was patching holes in the cob walls made by bees.

Tiny but functiona[l] kitchen. Width of walls determined b[y] cook's armspan: 5'[?]

"Mud Dauber," a 300 sq. ft. south-facing, earth-bermed passive solar cob structure. It has a "living roof" (plants growing in soil) and an earthen floor. When the site was excavated, the topsoil was separated and put on the roof. The rest of the soil was used to build the cob walls (which hold up the roof — making the cycle complete).

FAMILY HOMESTEAD IN TENNESSEE WOODS

In 1971, Johnny and Carol Kimmons and their family moved onto a 300-acre piece of forested land in the Sequatchie Valley in Tennessee. Over the past 30 years they have built a number of structures and have been living ". . . a sustainable lifestyle deeply integrated with the forest ecosystem." In 1996 the family homestead became a learning center and model for sustainable living and is known as the Sequatchie Valley Institute.

)))) www.svionline.org

A huge list of books on natural building materials:
http://store.yahoo.com/dirtcheapbuilderbooks/ind.html

Pressed flowers between panes of glass

Salvaged granite floor made of cut-off scraps from expensive kitchen counter tops

Greenhouse kitchenette with tub. Framing is local chainsaw-milled black locust (rot-resistant).

Cob "contra-flow" thermal mass heater with 10-foot cob chimney

Free-form spiral staircase of black locust made with chainsaw

"Como se Llama," llama barn and art gallery. Salvaged plexiglass scraps on frame serve as gallery shelves.

"Annoli Pajoli" post and beam structure was built from trees knocked down by an ice storm. Upstairs is a meeting and sleeping space; downstairs is a ceramics and glass studio.

KELLY AND ROSANA HART'S EARTHBAG-PAPERCRETE HOUSE

Kelly Hart

Kelly and Rosana Hart live in Colorado in a home-built earthbag/papercrete house. The Hart's website (www.greenhomebuilding.com) has a tremendous range of information about sustainable architecture and natural building materials: adobe, straw bale, cob, cordwood, earthbags, papercrete, cast earth, and lightweight concrete. Here are Kelly's photos and a description of building their home.

BUILDING WITH EARTHBAGS (sometimes called sandbags) is both old and new. Sandbags have long been used, particularly by the military for creating strong, protective barriers, or for flood control. The same reasons that make them useful for these applications carry over to creating housing: the walls are massive and substantial, they resist all kinds of severe weather (or even bullets and bombs), and they can be erected simply and quickly with readily available components. Burlap bags were traditionally used for this purpose, and they work fine until they eventually rot. Newer polypropylene bags have superior strength and durability, as long as they are kept away from too much sunlight. For permanent housing, the bags should be covered with some kind of plaster for protection.

There has been a resurgence of interest in earthbag building since architect Nadir Khalili, of the Cal-Earth Institute, began experimenting with bags of adobe soil as building blocks for creating domes, vaults, and arches. Khalili was familiar with Middle Eastern architecture and the use of adobe bricks in building these forms, so it was natural for him to imagine building in this way. The Cal-Earth Institute has been training people with his particular techniques, and now the whole field has expanded considerably with further experimentation by his students and others.

I have taken Khalili's ideas of building with earthbags that are laid in courses with barbed wire between them, and come up with some hybrid concepts that have proven to make viable housing. Instead of filling the bags with adobe

soil, I have used crushed volcanic rock. This creates a very well insulated wall (about as good as straw bale) that will never rot or be damaged by moisture. As a covering for the earthbags I used papercrete. This seems to be a very good solution to the need to seal the bags from the sun and the weather, without necessarily creating a vapor barrier . . . the walls remain breathable. Also, I expect the papercrete finish to be fairly maintenance-free, unlike an adobe finish that would require regular maintenance.

This is our first experimental earthbag dome. The interior diameter is 14 feet and the dome stands about 16 feet high. At first we tried filling the bags with the fine sand that it is built upon, but when we were partly done, the dome fell in because the sand couldn't hold the shape. Then we filled the bags with crushed volcanic rock (scoria) that provides better insulation and holds its shape much better. The arch over the doorway was created with a wooden form that was later removed. We kept the dome tarped most of the time until we papercreted the exterior, in order to keep the sunlight off the bags because the UV will eventually destroy them.

 www.greenhomebuilding.com

This is the beginning of the large elliptical dome that became our kitchen and living room. It measures approximately 30 feet by 20 feet. Because we are building on sand with excellent drainage and no problem of frost upheaval, there is no foundation other than a pad of 6 to 8 inches of the crushed volcanic rock (scoria). You can see the pile of scoria in the background, and a large wagon wheel in the foreground that will be used to support a circular window opening.

Here is the same dome as above, with joists in place for the loft and the arch form still supporting the entrance arch. The joists are simply resting on the bags and blocked up where necessary to maintain the level. Bags are then stacked between the joists and on top of them to lock them into place. Having the loft there made the structure much more sturdy as I continued to build. Two strands of four-point barbed wire were placed between each course of bags to help hold them in place and to withstand any tendency for the dome to bulge outward with pressure from above. We also placed a piece of baling twine under each bag which would be tied around three bags eventually. This provided more structural integrity and created a positive grip for any final plaster material.

Because of the elliptical shape, this dome required a rigid pole framework to help support the second story. I would not recommend building anything but a circular dome after this experience, because otherwise the forces are just not balanced enough. You see the large arch form for the six-foot-wide doorway. The house is a passive solar design, so we needed large openings to let in the sunlight. After several failures and much experimenting, we devised a double-bag technique to create such a large arch. Double side-by-side bags are used for columns at every doorway in the house.

On the left is the 16-foot interior diameter bedroom dome, and on the right is part of the large dome. Between them is the connecting portion of the house under construction. The back (north) bag wall is a section of a sphere that is braced into place with the rafters for the southern roof/wall. Other braces within the attic space help hold the shape.

Vaulted main entry to our house. Bell tower is on the left, and the bermed north side of the house, with completely covered pantry mound, is on the right.

Kelly and Rosana Hart

This shows the main entrance onto a landing, with the option of going up to the loft or down to the main level. Lots of natural wood was used to finish the interior components. An old woodstove for back-up heat is visible in the foreground.

Here I am applying a coating of papercrete to the outside of the large dome. I did this as soon as I could to protect the bags. Thermal pane glass was embedded in the papercrete on the outside over all of the circular windows.

▲ This is the papercrete tow mixer that was used to mix most of the papercrete. An invention of Mike McCain, the tow mixer is an amazing machine. It is made from a car rear end, a metal stock tank, a lawnmower blade and a few other parts. To make the papercrete, water is filled to within about six inches of the top, sand is added if desired, dry paper of virtually any description is added, and one bag of portland cement thrown in. One slow trip driving around the block produces a thick slurry that is total mush. This is drained through a sieve to eliminate the excess water, and then applied to the building. One mixer load yields between three and four wheelbarrows full of papercrete.

▲ This is the view from the landing down into the living room. One of our dogs is standing on the flagstone set into the adobe floor. The rest of the floor in the large dome is poured adobe that was scored with a rocklike pattern. This is a classic passive solar arrangement, with lots of south-facing glass and dark-colored thermal mass on the floor to absorb the heat. A window seat can be seen behind the dog, under the wagon wheel window. This seat was formed during construction with earthbags.

EMPLEO DEL CARTABON

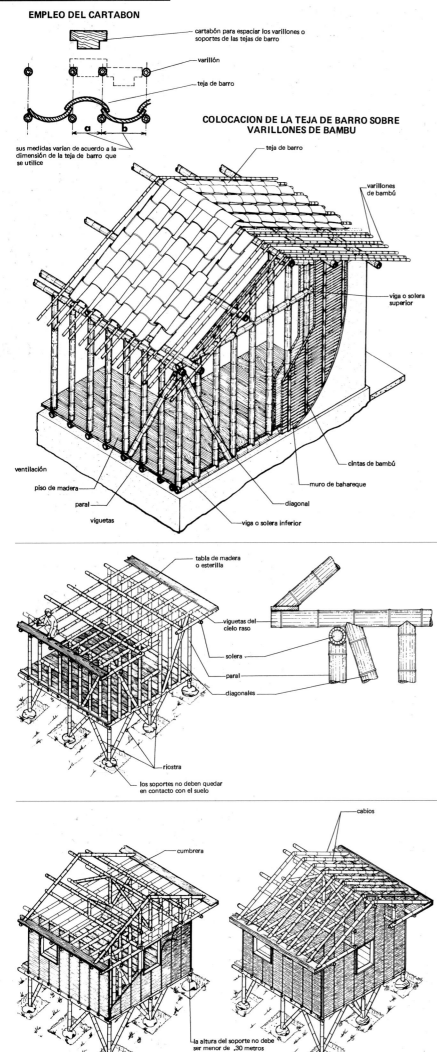

cartabón para espaciar los varillones o soportes de las tejas de barro

varillón

teja de barro

sus medidas varian de acuerdo a la dimensión de la teja de barro que se utilice

a b

COLOCACION DE LA TEJA DE BARRO SOBRE VARILLONES DE BAMBU

teja de barro

varillones de bambú

viga o solera superior

cintas de bambú

muro de bahareque

diagonal

viga o solera inferior

ventilación

piso de madera

paral

viguetas

tabla de madera o esterilla

viguetas del cielo raso

solera

paral

diagonales

riostra

los soportes no deben quedar en contacto con el suelo

cabios

cumbrera

la altura del soporte no debe ser menor de .30 metros

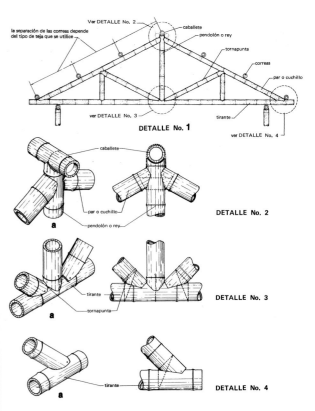

DETALLE No. 1

caballete

pendolón o rey

tornapunta

correas

par o cuchillo

tirante

la separación de las correas depende del tipo de teja que se utilice

DETALLE No. 2

DETALLE No. 3

DETALLE No. 4

tirante

tornapunta

BAMBOO: GIFT OF THE GODS

Oscar Hidalgo-Lopez

Oscar Hidalgo-Lopez has had a love affair with bamboo for over 50 years. An architect *(Universidad Nacional de Colombia)*, he has travelled all over the world giving lectures and conducting workshops on bamboo construction.

He built his first bamboo building in 1965, a kiosk at a country club in Bógota, Colombia, and later was a UN adviser on bamboo construction in Ecuador, Colombia, Bolivia, Peru, and Costa Rica. In 1982–85, he built 12 houses in Ecuador. In Colombia he built three houses, was construction consultant in the building of 100 bamboo-framed houses, and had a small factory making bamboo furniture.

His 1974 book *Bambu* (in Spanish) is a classic on bamboo construction. In 2003, Oscar published a 550-page encyclopedia on bamboo construction (in English), *Bamboo — The Gift of the Gods*. It's a must for anyone interested in building with bamboo.

You can order the book from the website below or by emailing Oscar direct.

www.bamboodirect.com
bamboscar@email.com

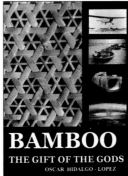

BAMBOO
THE GIFT OF THE GODS
OSCAR HIDALGO - LOPEZ

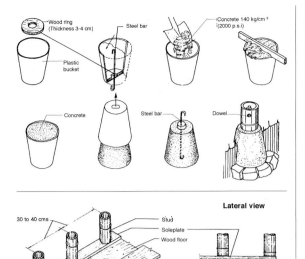

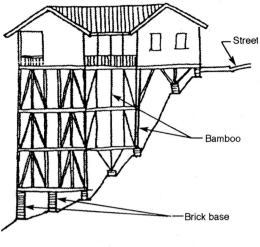

Lateral view

30 to 40 cms

Stud
Soleplate
Wood floor
Beam
Wood joists

30 to 40 cms

Street

Bamboo

Brick base

In Manizales and other towns and rural areas of Caldas state in Colombia, the people used to build their houses beside roads or streets, even though they had to build a substructure 15 meters high (4 to 6 floors) to support the house which generally had one or two floors. The construction of the substructure was very common up to the sixties because bamboo had no value.

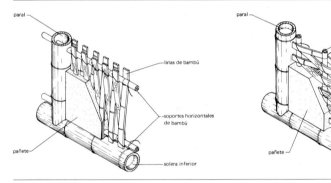

paral
latas de bambú
soportes horizontales de bambú
pañete
solera inferior

paral
soportes verticales de bambú
latas de bambú
solera inferior
pañete

Traditional Campesino Structure — 12m diameter
Bamboo slats nailed on all around for tensile strength

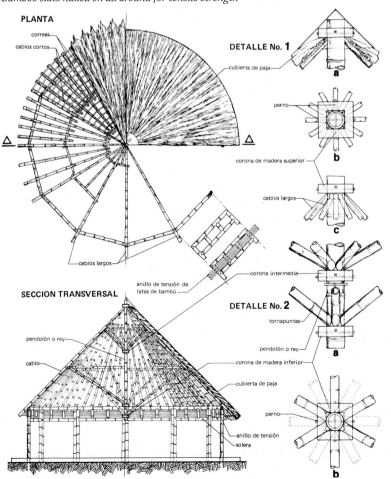

PLANTA

correas
cabios cortos

cabios largos

SECCION TRANSVERSAL

pendolón o rey
cabio
cubierta de paja
anillo de tensión
solera

DETALLE No. 1
cubierta de paja
a

perno
corona de madera superior
b

cabios largos
c

corona intermedia

DETALLE No. 2
tornapuntas
pendolón o rey
corona de madera inferior
cubierta de paja
perno
anillo de tensión de latas de bambú
a

b

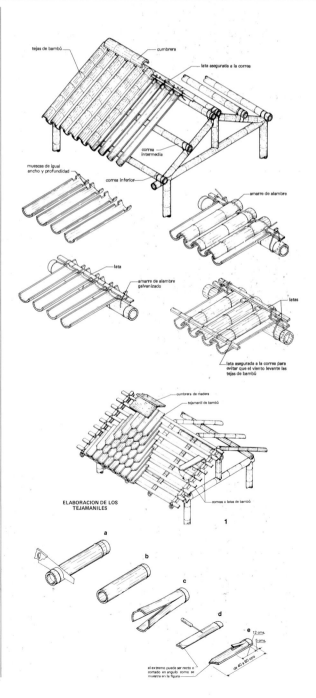

tejas de bambú
cumbrera
lata asegurada a la correa
correa intermedia
muescas de igual ancho y profundidad
correa inferior
amarre de alambre
lata
amarre de alambre galvanizado
latas
lata asegurada a la correa para evitar que el viento levante las tejas de bambú

cumbrera de madera
tejamanil de bambú
correas o latas de bambú

ELABORACION DE LOS TEJAMANILES
1

a
b
c
d
e 12 cms.
5 cms.
de 40 a 60 cms.
el extremo puede ser recto o cortado en angulo como se muestra en la figura

more...

Typical Bamboo Joinery

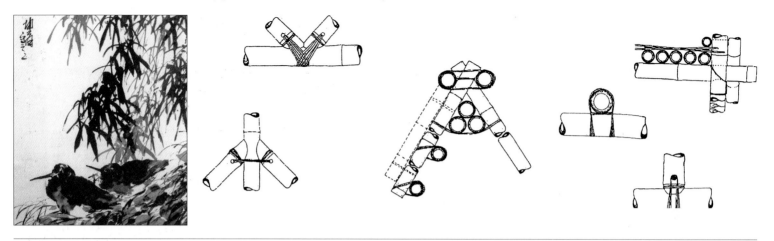

Bamboo Bridges

ARMADA DE LA UNIDAD

FIJACION DEL CABALLETE

nudo de levita que permite zafar el soporte desde la orilla

UNIDAD

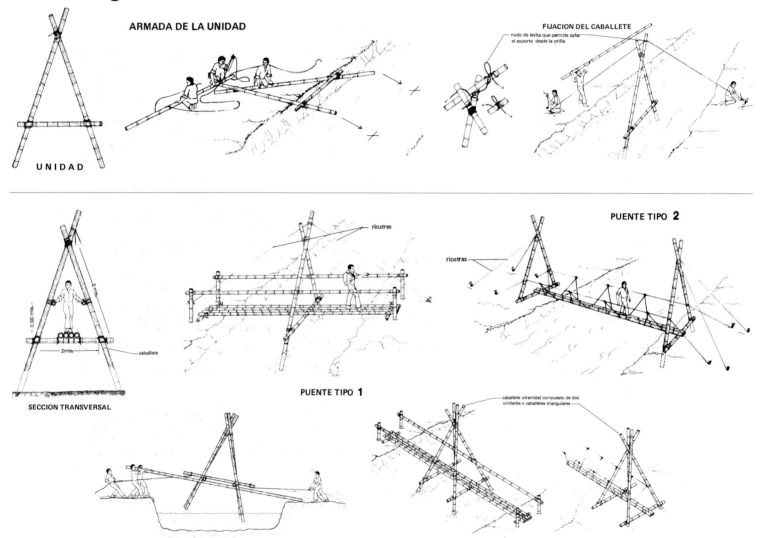

riostras

PUENTE TIPO **2**

riostras

3 mts.

0.90 mts.

2mts.

caballete

SECCION TRANSVERSAL

PUENTE TIPO **1**

caballete piramidal compuesto de dos unidades o caballetes triangulares

Manufacture of Large Water Tanks with Bamboo

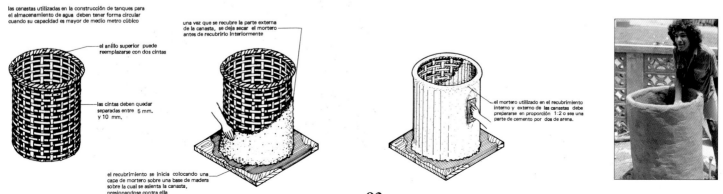

las canastas utilizadas en la construcción de tanques para el almacenamiento de agua deben tener forma circular cuando su capacidad es mayor de medio metro cúbico

el anillo superior puede reemplazarse con dos cintas

las cintas deben quedar separadas entre 5 mm. y 10 mm.

una vez que se recubre la parte externa de la canasta, se deja secar el mortero antes de recubrirlo interiormente

el recubrimiento se inicia colocando una capa de mortero sobre una base de madera sobre la cual se asienta la canasta, presionandose contra ella

el mortero utilizado en el recubrimiento interno y externo de las canastas debe prepararse en proporción 1:2 o sea una parte de cemento por dos de arena.

Ulloco's Bridge with Roof Cover, Colombia

Basic Wooden Bridge Types That Can Be Built with Bamboo

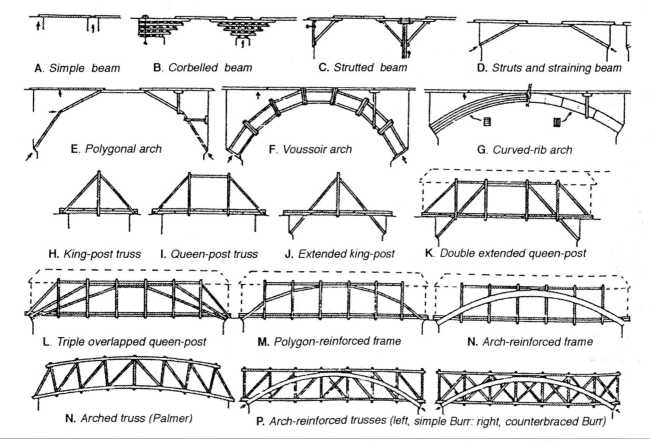

A. Simple beam **B.** Corbelled beam **C.** Strutted beam **D.** Struts and straining beam

E. Polygonal arch **F.** Voussoir arch **G.** Curved-rib arch

H. King-post truss **I.** Queen-post truss **J.** Extended king-post **K.** Double extended queen-post

L. Triple overlapped queen-post **M.** Polygon-reinforced frame **N.** Arch-reinforced frame

N. Arched truss (Palmer) **P.** Arch-reinforced trusses (left, simple Burr: right, counterbraced Burr)

Bamboo Windmills

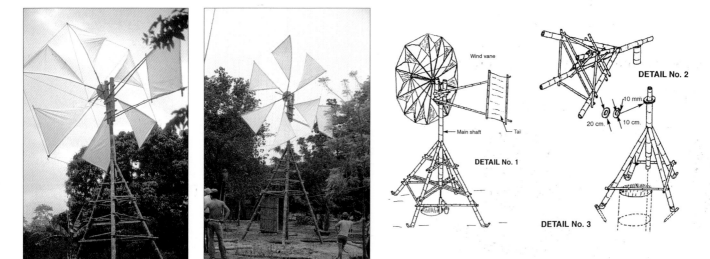

RAND AND COOKIE'S LOG HOUSE

Rand Loftness and Cookie Brown

IN 1977 my lady and I decided we were tired of paying rent and that we needed a home of our own We had a copy of *Shelter*, and although we knew nothing about building or designing, somehow the book made us believe we could do it. We pored over it until it was just a pile of unbound dog-eared pages; it encouraged us to rely on instinct, imagination and native ability, rather than money and phone calls, to make our house. That was just as well, as we had little money and no phone. Now after twelve years of making it up as we go along, we have a house that's not only paid for, but is the focus of our lives.

As it happened the first piece of land we looked at fronted on a small inlet of Puget Sound and had a view of the Olympic Mountains. Not only that, but we could just barely afford it. So we bought it. It was five acres of jungle with some nice trees, a large flat area and a steep 200-foot drop to the bay. We jumped right in, clearing brush with machetes and talking all the while about what our house would be like: Would 40 × 60 be big enough? Oh probably. How

about two stories? It has to have an upstairs, don't you think? Well, of course, and a basement too. And it has to be a log house, doesn't it? Absolutely.

We soon settled on the most practical spot to build — at the top of the hill with a view to the north and sun from the south. Next we began looking for logs and we bought a truckload of turned poles for about $2000. That was a third of the six thousand we had borrowed to build with. For another three hundred a man came out with a small Cat and dug a hole for the basement.

"You kids gonna build this house yourselves?" he asked us.

"Yep."

"Well, you know how a mouse eats an elephant, don't you?"

"No, how's that?"

"A little bit at a time," he answered with a smile and a wink.

Over the years it's been good to keep that in mind, along with one other piece of advice from an older guy: keep it simple.

When the hole was dug we found that without too much extra hand digging we could fit in

Cookie in her rainsuit on the first truss

the footing for a twenty-eight-foot square. So we went looking for an old cement mixer. At the time we had a '55 Cadillac and as soon as we got serious about building we traded it for a '67 Ford pickup whose first duty was to go get the cement mixer and then haul water from town so we could mix concrete.

We figured we needed eight-foot walls in the basement so we counted up how many blocks that would be and ordered them up, along with the cement, sand and lime needed to mix mortar. That took another thousand. The money was half gone, but we were about to start really building. We had no idea how long things would take but we were wildly optimistic. All the while that we were laying blocks we were working at our jobs

and staying up late reading building books in the library. We were getting an idea of how we were going to use the logs and what the house was going to look like, but it was still pretty much designed on the back of an envelope as we went along.

With the logs that we had it appeared that there wouldn't be enough to overlap them in the corners in classic log-cabin style. So we decided to use four of them as corner posts and lay the logs in between them. Milled flat on two sides with a chainsaw mill they provided a good enough surface to nail to, with big foot-long nails in pre-drilled holes. As we laid the logs on top of one another we ran the chainsaw between them until it cut both logs the whole length, and this made a pretty good fit and kept us busy for a few months.

We had already decided that for the upstairs and the roof we would use the logs to frame with, though we didn't know how we would do it. At the time I didn't know a mortise from a tenon and thought that timber framing was how they made railroad bridges. But fortunately, just in time, we found some hints in a log building book that got us started on timber framing with round logs. We just drew up some shapes that we liked and used our high school trigonometry and a calculator to find the lengths of the pieces, locate the mortises and find the shapes of the mortises. Of course all that was pretty abstract until we set the first logs up in horses and fired up the chainsaw. We'd soon learned that it's not too easy to make a square cut on a round log, or to cut accurately with a chainsaw. And it took a real leap of faith in our numbers to cut the first mortise. So it was both a relief and a quiet sort of thrill to fit the first tenon into its mortise and call it a fit. After that it was

Woodshed ready for winter. Built on 4"-thick edge-supported structural slab with integral steps

Solar greenhouse brightens and heats house.

several weeks of careful work before it was time to round up our friends, buy some beer and have a raisin' day. That was the first of many parties in our house, the first time it really felt like it was going to be a house, and a hopeful sign of good times to come.

But when it was over there was still a great deal to do before we could move in and it had to be done pretty quickly for we couldn't keep paying rent and making land payments for long before the money ran out. So over the summer we had to close off the upstairs, make a front door, make some windows, get some electricity to the house, do something about water . . . the list seemed endless.

For a water system we had some plastic apple juice barrels, one of which we put on some logs protruding from the second story. The other we kept in the pickup and filled in town or at a local park. Then we could back the truck under the first barrel and fill it from the truck with an electric drill pump, and gravity feed it into the makeshift kitchen. That was fine until the weather got good and cold, five below for days on end; then it froze into a solid block that didn't thaw till spring.

We also had to make a stove to heat the house, and for that we relied on a stove-making book by Ole Wik, who appeared in *Shelter*. It turned out to be an upright 55-gallon drum with a 30-gallon drum inside of it, a downdraft stove with a real personality. We'd load it up with green wood, and it would cook it till it was dry and then burn hot and the stove would gasp for air with an increasingly frantic rhythm until it would blow the top clear across the room, usually in the middle of the night.

Nonetheless it was a good old stove and it kept the house livable until it burned through and we had to make another . . . of a slightly different design.

We decided that we had to move in at the end of September no matter what, so with a stove and a funky water system we threw some plastic over the roof and made a permanent camp. Our plan all along had been to make a big box and then figure out how to live in it. So when we moved in we had no walls to obstruct our thinking. We had no stairs either, not to the basement or the sleeping area on the upper floor. There was plastic in many of the window holes, no real way to bathe, no insulation, and none of a host of other things, but we didn't care. It was our house and we were going to make it work.

We've been here for thirteen years now, and a little bit at a time it's turning into a pretty civilized place. We found a small spring on the hill below the house and corralled it by digging it out and placing a perforated bucket in it surrounded by gravel. We led a pipe out of the bucket into a settling tank and then to a 1000-gallon tank further down the hill. We put a submersible pump in the big tank and that gave us a pressurized water system. So even though the spring is only a couple of gallons a minute, with the

Spring-fed, hand-dug, solar-heated swimming pool; greenhouse fiberglass over EMT conduit trusses provides heat.

storage tank, we have all the water we can use.

Having a water system encouraged us to think about making a septic tank which we laid up with four-inch concrete blocks plastered with mortar and coated with tar. So now after ten years or so with the composting toilet, which really worked OK, we have a flush toilet and a shower.

Since we found the downstairs a little gloomy with the solid log walls, and were perched on a good solar site, we cut out most of the south wall and built an attached solar greenhouse with a concrete floor. Suddenly it was transformed into a bright and cheerful place that was much easier to heat. And of course there's plenty of room for plants, and that livens up the whole house.

WE LITTLE REALIZED when we began here that we were setting into motion a process that would become our whole lives. Our education, skills and thinking have evolved along with the place in a way that causes one thing to lead to another and then another and then another. We've discovered that we can do about anything that we can think of, and anything we don't know is somewhere in the library.

It is somewhat of a paradox that the intent of this consumer society is to make people more and more dependent on the services it provides, which one must submit to wage slavery to pay for, while at the same time the information that makes self-reliance possible is available for free in the library.

From the beginning one of the most basic facts of our lives has been that there isn't much money, so we have constantly had to think of how we could improve our lives without spending any. Naturally we soon became passionate scroungers, to the point even of

going to the ocean and dragging planks out of the surf when we couldn't afford to buy any at the lumber store. Even now when things are pretty squared away and we're no longer really broke, we're still bubbling with subversive glee at being able to build what we want with little regard for the "normal practices" of the main consumer building industry.

Cookie and Rand

Of course there's more to making a place than just building. By now we've beaten back enough of the jungle to have a garden and fruit trees and a fenced area for pigs and chickens, and are about to start on some fish ponds for carp and trout, using the overflow from our springs. We're turning into regular peasants, which is what we always really wanted. It's not a vocation that commands much prestige, and you don't hear a lot about it from the career counselors, but once you learn to think for yourself and do for yourself, it's a lot of fun.

The *Shelter* book meant so much to us that we're glad to share our experiences. The world would be so much better if people would learn to do things for themselves instead of being intimidated by the "professionals," government regulations and the prevailing mindset of our so-called culture into thinking that the only way to have a house is to be a wage slave until you can afford to buy one — a vain hope at today's prices.

more...

SCISSORS TRUSS FRAME

$\mathbf{A}$LL ALONG the main question has always been, how can we improve our lives without spending any money? When the time came to make outbuildings, this question eventually directed my attention to the plentiful supply of small fir poles, commonly called "pecker poles" available in the local forests. I figured that by joining these poles with mortise and tenon joints, I could make a building for practically nothing. So I started with a woodshed and succeeded well enough that I began to see other possibilities.

The accompanying photos show the result of my most recent effort, a 20 × 60 building we call the motor stables. Though tricky to make, these scissors trusses that are the crux of this building use less material than king post trusses and they are quite rigid laterally, so they are easy to raise without damage. This building cost about 100 dollars to get as far as the space sheathing: forty dollars for the concrete in the piers and sixty for most of the poles, from a local firewood yard. The rest of the poles were collected on a permit from Simpson Timber. Once you get to the roof, of course, you have to spend some money. For small buildings like woodsheds I've managed to scrounge enough cedar to do it for free, but usually this isn't possible.

–Rand Loftness

PHOTOGRAPHERS

Small barn in Yamanashi Prefecture, Japan was designed by Shuhei Hasado, and is used for vegetable storage. The framing is tree branches, which were then plastered, and embedded with old barn shingles, and capped with rice straw.

THE PHOTOGRAPHS OF YOSHIO KOMATSU

Drawings by Eiko Komatsu

I LOVE BOOKS on building. In bookstores, people's homes, at book expos, I look for books on homes, dwellings, family-sized structures as opposed to the monumental edifices or architectural homes for the wealthy. After 30+ years of pursuing this passion, I came across a book that just stunned me. It was called *Living on Earth,* by Yoshio Komatsu (Fukuinkan-Shoten Publishers, Tokyo, 1999) and it was a masterpiece. 1700 meticulous color photos of people's dwellings all over the world, built from natural materials. It was in Japanese, but that didn't matter. The photos spoke for themselves. Wow!

Since 1985, Yoshio has travelled all over the world shooting photos of vernacular homes. Eiko often goes with him. He shoots with two Canon EOS cameras, mostly using two Canon "L" zoom lenses, a 16–35 mm and a 28–70 mm. The buildings he finds are wonderful and unique. Moreover, the people in the photos look relaxed and happy — obviously comfortable with the photographer. The American version of this book — *Built by Hand* — and edited by Athena and Bill Steen, has just been published by Gibbs Smith *(see p. 234).* Here are eight pages of photographer-extraordinaire Yoshio Komatsu's work, with drawings by Eiko Komatsu:

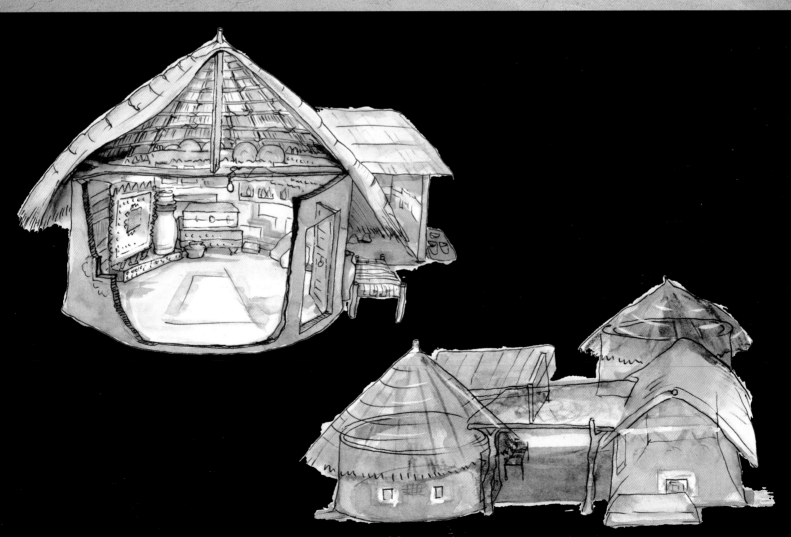

INDIA This immaculate dwelling is in the village Ludia, in the Kutch region Gujarat near the Pakistan border. The walls are built with earth, straw, and cow dung and decorated with beautiful patterns. The interior walls are plastered and have small embedded mirrors.

more...

MONGOLIA *Yurt ("ger") in a plain 150 miles south of Ulaan Baatar. Mongolian nomads move their yurts as they follow their grazing stock. The yurts can be assembled in three hours. The wall is a wooden lattice work that expands for erection and contracts for travel; the roof is supported by poles that connect at the center ring. The outer skin consists of thick wool felt mats. In the center is a horse dung-fired cookstove.*

TOGO *The stunning house (shown at right) appears to grow out of the red soil. It is a series of towers connected by thick walls, an earthen castle. The rooms under the capped conical roofs are for storage of millet and sorghum; the other upper story tower rooms are for sleeping and cooking. The ground floor is for animals. On the exterior walls are fetish animal skulls and peep holes for spying on potential enemies. Note family of ducks at right heading for their hole in the wall.*

Yurt under construction

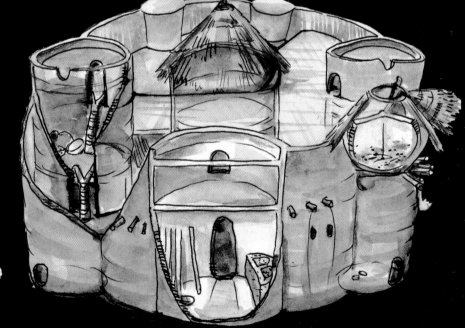

1. Workshop & entrance
2. Bedrooms
3. Kitchen
4. Granary
5. Chicken's room
6. Peeping hole

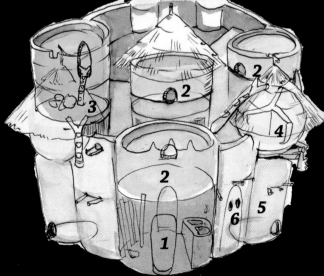

more...

INDONESIA Houses of the Bajau people stand on the ocean's edge in Manado, northern Sulawesi Island. They are fishermen and grow and sell seaweed. The simple house consists of a mangrove floor, bamboo walls, and a palm leaf roof.

SENEGAL Fadiout is 70 miles south of Dakar. The village is on a small shell-mound island and reached by a wooden bridge. There are basket-like granaries on stilts outside the village; these were built over the water after a big fire many years ago, to protect stored food.

VIETNAM Floating homes on Langa Lake, 80 miles northeast of Ho Chi Minh City. These people live on the water and raise fish and alligators in water cages that they sell in the marketplace.

MALI Houseboat on Niger River near Mopti. Mopti is at the junction of the Niger and Bani rivers, an active trading place in the Sahara Desert zone for a great variety of people. The Bozo people shown here live on and fish from this boat; they also have straw huts along the river for drying the fish they sell on the market.

MYANMAR (BURMA)
Inle Lake

BENIN *Lake Nokwé*

VENEZUELA *Above and below: The Orinoco Delta is inhabited by the Warao Indians. Their stilt houses, built along the river banks, have no walls. The importance of the river in their lives is reflected in their name. "Wa" in their language means "canoe," and "arao" means "people."*

PAPUA, NEW GUINEA *Spirit house*

more...

VENEZUELA A shabono, a circular structure about 100 feet in diameter inhabited by the reclusive Yanomami Indians in a remote part of the Amazon jungle, upstream of the Orinoco region, near the Brazilian border. The shabono actually consists of a series of individual dwellings side by side, each with its own hearth, grouped together for defensive reasons. Between 50–70 Yanomami, who subsist entirely from the forest, live in one of these structures. Photo: Yoshio Komatsu

Timber and mud-chinked farm house near the village of Nagar in the Himmalayan valley of Kulu. Animals live below, people above, with a balcony for the view.

A brick caravansary in Herat, Afghanistan is now used as a storage depot for a coppersmith.

KEVIN KELLY

Above and below: Slate-roofed houses, Kulu Valley, Indian Himalayas

KEVIN KELLY travelled throughout Asia in the '70s and '80s with a backpack and two cameras. He visited Japan, Burma, Thailand, Sri Lanka, Nepal, Iran, Afghanistan, Pakistan, and India and shot some 40,000 pictures. In 2002 he selected 600 of these and published *Asia Grace*, a rich and diverse mix of Kevin's experiences. The book is unique in that there are no captions — just wall-to-wall photos.

Kevin is a man of many talents, probably best known as executive editor of *Wired* magazine in its exciting early years. Prior to that he was publisher and editor of the *Whole Earth Review* and oversaw publication of four versions of the *Whole Earth Catalog*. In 1981 he started *Walking Journal*, the first magazine on walking. He was a founding member of The WELL, a pioneering online newsgroup, the author of several books, a writer whose work has appeared in *The New York Times*, *The Economist*, and *Esquire* and is currently, along with Stewart Brand and Ryan Phelan, working on the All Species Inventory, a plan to document all living organisms in the span of one generation.

On these five pages are photos from Kevin's seven wandering years in various parts of Asia. At the websites listed below you can see the entire book online, as well as Kevin's past and present interests.

My method of shooting was simple: shoot first and ask questions later . . . I traveled solo most of the time. I had lots of time and no money . . . I would leave the U.S. with 500 rolls of film in my backpack . . . The majority of the images were shot using a pair of Nikkormat camera bodies . . . and five very heavy, old-fashioned lenses . . .

www.asiagrace.com
www.kk.org

On the rugged peninsula of Mount Athos, Greece, monks have built anchorites and hermitages along the coast.

The Mogul influence is felt in this Rajastan Palace situated on an island in a lake, and now run as a small hotel.

One of many tent structures built to house the largest gathering of humans on earth: 14 million pilgrims during the 1976 Khum Mela in Alahabad, India.

Peaks have always been holy places. In the eastern mountains of South Korea a woman devotee worships at an altar built into rock cairns. She has lit candles and is waving a Korean flag.

Floating homes fill Aberdeen Harbor on the backside of Hong Kong. Residents use water taxis to commute to shore, while floating stores peddle goods to the waterborne community, and floating restaurants serve meals. Some of the houseboats can sail, but most homes are fixed-up barges that need to be towed.

more…

SIKKIM *Young monks*

Photo: Kevin Kelly

HIMALAYAS, NEPAL *These narrow terraces in the hills of the Himalayas (only five feet wide in places) are for barley, oats, and millet. They require tremendous upkeep against erosion and gravity. The hut at center is a day-use shelter, not a home.*   Photo: Kevin Kelly

more...

Temple island in Bali
Photo: Kevin Kelly

THE HALLIG HOMES OF NORTHERN GERMANY

Hans Joachim Kürtz

Houses on Hallig Langness. Today all the mounded building sites on the Halligs ("Warften") are protected by their own dikes, as shown here.

Hallig Houses, as shown in these three photos, have been built though the centuries either singly, or in small groups on "Warften," artificial mounds built up above sea level.

Next two pages (114-115):

Hallig Habel during "Land unter," a local term describing the flooding of the Halligs during storms when just the houses stick out of the water. Twenty years ago, when this picture was taken, the house was inhabited by a farmer. His sheep and cattle spent their nights in the lower story. In extreme storms, when the lower story was flooded, the farmer would bring his animals upstairs.

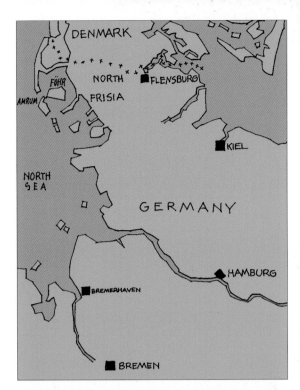

Hallig Süderoog during "Land unter," looking much like a ship at sea. Like Hallig Habel, Süderoog is one of the smallest inhabited Halligen, with just one house. The circular white fence at left surrounds the "Fething," a pool lined with straw and clay that collects drinking water. (Today most of the Halligs have a water connection by pipes to the mainland.)

Hallig Hooge

Hallig Langness

Hamburger Hallig

The Hallig postman of Langness. He delivers mail via a motor-powered lorry on a tiny railway line from one Hallig to another. The tracks are dry during low tides, underwater at flood tides.

One of the farmhouses of Hallig Hooge ("Land unter"). During storm floods, the North Sea separates neighbors from each other, as shown here.

I<small>N THE</small> <small>NORTHERNMOST</small> part of Germany, just below the border with Denmark (about 60 miles northwest of Hamburg), is an area on the North Sea known as North Frisia. It is an area of low flatlands and mudflats, with tidal changes of up to nine feet, so that at low tide, North Frisia's surface area is one-third larger.

Just off the coast is an archipelago of 14 small islands (the largest four being Sylt, Föhr, Amrun, and Pellworm). The other ten islands are "Halligs," small, low areas that are a cross between an island and a sandbank, and are unique in the world. Several hundred people live in houses standing on built-up mounds known as "Warften."

About 30 times a year, high tides completely surround the Warften, isolating them from other Warften on the same Hallig, so that from the air, the archipelago appears to consist of some 50, rather than 14 islands.

In times past, the North Sea (nicknamed "Mordsee," or Killer Sea), would batter the area. In 1362 and 1634, huge winter storms caused the "Grote Mandränken," or "Great Drownings." After heavy storm floods in 1962, dikes about 20 feet high were built around the houses on the Halligen, so they are no longer flooded.

more…

LIKE A GIANT MANDALA, the island village of Mexcaltitán seems to float on the Rio San Pedro delta (between Mazatlán and San Blas). This photo was shot in late summer 1968, when torrential rains turned streets into canals. At lower right is a partially submerged basketball court; U-shaped building at lower left is fishermen's cooperative. Photo: W. E. Garrett, National Geographic Society

Houseboat on the River Kwai,
Damnoen Sudak, Thailand
Photo: Robert Barab

Monastery in Meteora, Greece
Photo: Clay Perry

Waterside restaurant in Kekova, a Turkish island in the Mediterranean Sea
Photo: Dr. Mehmet Hengirmen

FANTASY

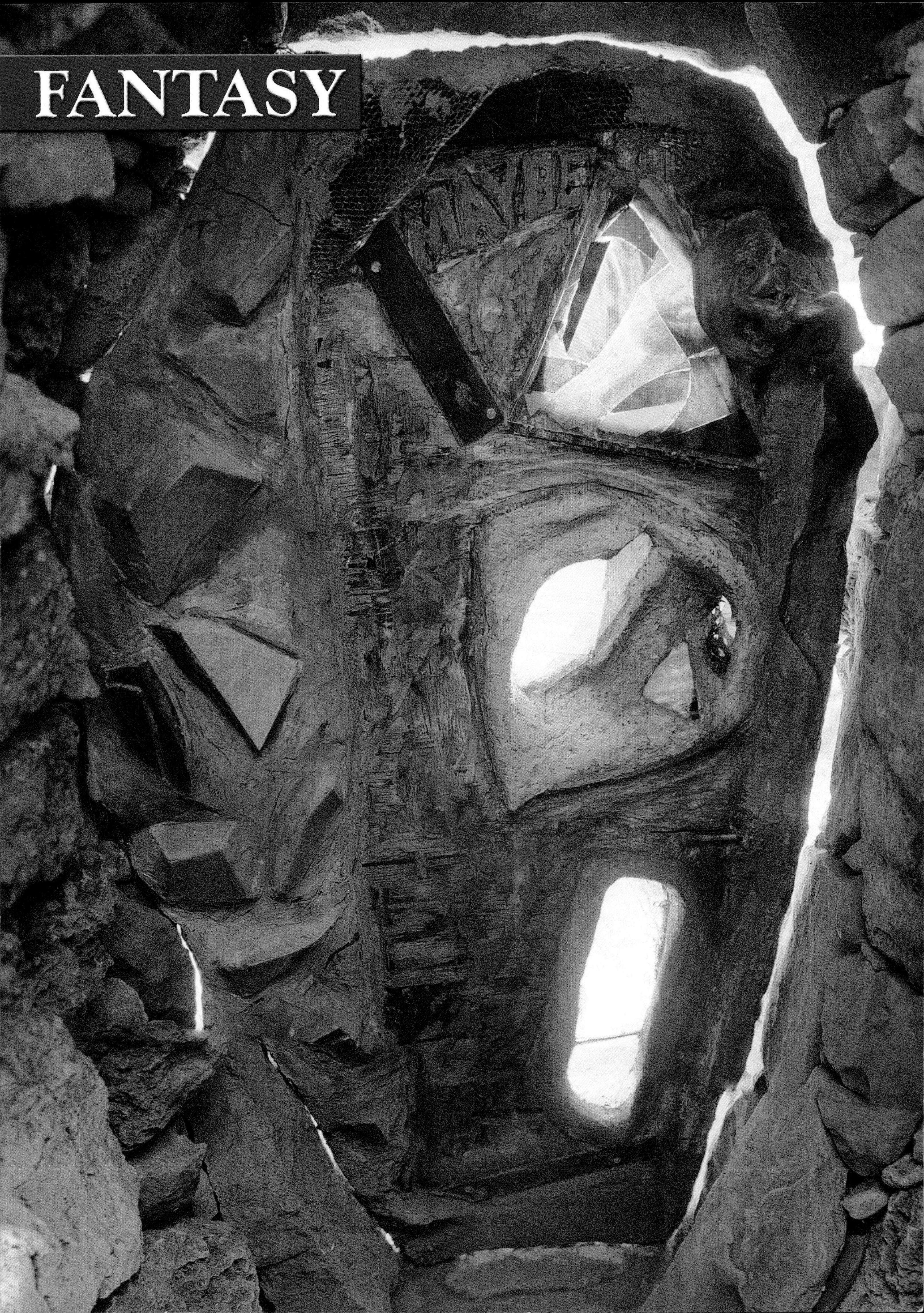

ELIPHANTE
Michael Kahn's Sculptural Village in the Arizona Desert

> *"When we started to build living space, we decided to make it an art form."*

MY COUSIN MIKE started painting when he was 12, and he's been an artist most of his life. No compromises, no job that would interfere with his art. We're a year apart, sons of two brothers, and we played together as kids. We both went off to college in the '50s, and I lost track of him until the early '60s, when we lived next to each other for a year in Mill Valley, California. Then in 1965, on one of those "consciousness-expanding" road trips of the times, I visited him in Provincetown, Cape Cod. Since then we've kept in touch over the years. So when I look at Mike's work, I have a 60+ year perspective, and am hardly unbiased. That being said, I think Mike has created a major American work of art, conceived, built, and lived in, and unknown to the "art" world. Mike's a shy guy, and puts his efforts into his art rather than marketing. People don't know about him. He doesn't get grants from rich people or big corporations. This is

why it's exciting to show you his work in these pages.

Mike graduated from the University of California at Santa Barbara in 1958. He intended to go to graduate school to become a school psychologist, but met a portrait painter in New Orleans who rekindled his interest in painting. He moved to New York to study at the Art Student's League and National Academy, then eventually to Provincetown where he studied with Henry Hensche. He then

moved to Crete, living in an old farmhouse with a Mediterranean view and worked on a series of oil paintings. His wife Leda joined him there and they eventually moved to Paris, then back to Provincetown.

They spent a year building a camper on the back of a 1960 Ford flatbed. "It looked like a covered wagon." Mike felt the urge to do large paintings at the time and had read Max Ernst's description of the landscape around Sedona, Arizona, ". . . the forms of red rocks the closest to

inner vision." They drove the truck to Sedona in 1977 and met a man who told them they could set up camp on three acres of land along the banks of a river near Cornville. They started out with a leaky tent and the camper/truck. They had a little shed filled with their books (the "Winter Palace"), and cooked at an outdoor kitchen under the cottonwood trees.

"When we started to build living space, we decided to make it an art form."

more...

Greenhouse room built out of old auto windshields siliconed together. Stained glass is siliconed onto inside of glass.

more...

South wall (a favorite hummingbird site)

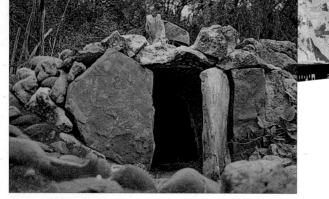

Kiva. Ladder with rock door goes down 8 feet to cool and quiet carpet-covered meditation space.

Eᴌɪᴘʜᴀɴᴛᴇ, the first major building, was constructed in three years. It was built out of ". . . adobe, wood, stone, ferro-cement and glass." To enter, you wind through a sculptural painted tunnel into a room with stone floors, steps, terraces, and a multi-colored profusion of stained glass. The center of the large glass wall ended up looking (unintentionally) like an elephant. A friend came by one day, looked at it, and said "Eliphante," and so the building was named. There is a large pond outside Eliphante, a solar-heated shower building, an underground kiva, and sculptures of rock, mirrors, and found objects throughout the grounds.

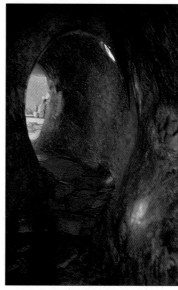

Looking in from the entrance tunnel

*Side view of room shown on pp. 124-25
(and in right corner of photo at left)*

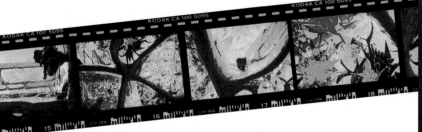

*North wall with piano and wood collage of carved 2x4s;
door in collage opens into secret back room.*

Mike and Leda in their outdoor (summertime) kitchen

Bath house

Outside door

PIPE DREAMS

PIPE DREAMS is Mike's most recent building, and it's an art gallery of his paintings, a labyrinth of a building with dozens of rooms, tableaus, displays of paintings, fabrics, tilework, stonework, stained-glass-colored beams of light reflecting on walls and floors. Mike's friends David O'Keefe and Michael Glastonbury collaborated with him in the spontaneous creation and construction process.

As you can see from these eight pages, Mike ended up building a sculptural village. He says he was inspired mostly by sculptor Conrad Malicoat of Provincetown, and by the book *Shelter*, ". . . not only ideas and techniques, but an undercurrent of creative inspiration, a trust that things will work out."

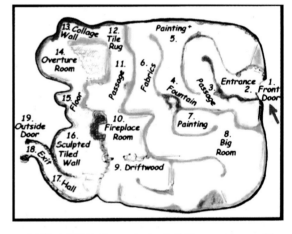

Collage wall by Michael Kahn and Michael Glastonbury

Inside of outside door

Wall of dry stacked rock and driftwood

WEB: www.eliphante.org
EMAIL: eliphante@verdenet.com

Front door

Oil painting (completed in 1970) on display in its own room

Painting on wall by Michael Glastonbury

Pipe Dreams namesake sculpture (on top of roof)

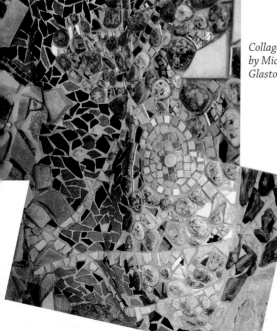

Collage wall by Michael Glastonbury

Painted rock floor section

MA PAGE'S BOTTLE HOUSE

JOHN AND MAXINE (Ma and Pa) Page moved away from the city (San Francisco Peninsula) in the '70s to a remote valley in the high desert country of Nevada. John took up mining, resurrecting a small placer mine, and Maxine set about to build a house out of bottles where she could store her paintings and art projects.

I ran across the bottle house early one morning while out looking for hot springs in the Nevada desert. The rising sun was shining through the bottles, it was this glowing multicolored apparition in a remote desert valley. Wow! In ensuing years I went back to visit and photograph the Pages and their projects, taking my kids along. Here are some photos, parts of a newspaper article, and an excerpt from one of Maxine's letters.

In the early '70s Maxine Page worked at Hewlett Packard in what has since become Silicon Valley. Mother of six, Maxine had been divorced several years earlier, and was living in the suburbs with her kids. One night she came home, ". . . and there was this young guy — John — sitting around with the kids. I thought he was a friend of my sons."

It turned out that John, a year younger than her oldest son, was interested in *her*. "He had been to Vietnam, was 25, and looked 35 and I was 45 and he told me I looked 35."

John and Maxine hit it off, despite the 19 years' difference in age. "We honky-tonked together." On August 5, 1974, John's 25th birthday, they were married in San Jose. Maxine was 44.

"You can imagine, I took a lot of ribbing at work. 'How's your *young* man?'"

They started going to Nevada and exploring ghost towns. In 1975 they found a ghost town — Fitting, Nevada, about 17 miles from Unionville — with nine buildings left. John met the county sheriff, who ended up telling the Pages they could live in one of the old buildings if they would be caretakers. They moved into a two-bedroom house that had been inhabited by Chinese working in the nearby Bonanza King Mine. There was an old orchard, with apple, cherry, apricot and pear trees, and John ran a PVC pipe to it from a creek and started watering the trees. In summer, locals would come out and pick the fruit. It was a good life.

John and Maxine lived for years with very few amenities, much like 19th-century miners. After about two years the mining company that owned mineral rights to the land demolished the town, and the Pages had to move on. Maxine and John then moved to Spring Valley, near Lovelock, where they leased 280 acres from the BLM. On the abandoned homestead they built a 12' × 12' cabin to live in, and started mining. "First we panned; we might get $20 worth of gold a day." Back then gold got to be about $800 an ounce. "You could put $20 worth of gold in a thimble."

Next they got a dry washer. "We got enough money from that to build a full operation." John got an antique (1910) bulldozer and then a 1939 Ford dump truck ("the same year John was born"). They got an Allis Chalmers loader and a used trummel, the device that breaks up the rock. The rock is loaded into a hopper and a belt moves it to the trummel. Any

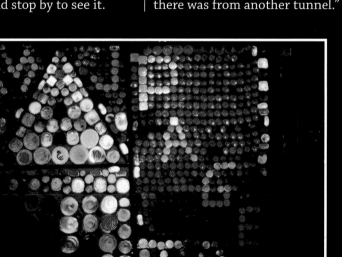

gold then falls out in the bottom of the catcher and is washed down in the riffles (grooves). On a good day they would get $100 worth of gold. John had been a welder in the Navy and had to constantly repair and fix the equipment.

There was a creek running through their land, fed by five artesian springs. "I had a garden. Carrots, squash, beans, potatoes, broccoli, lettuce . . . onions all around the outer edge . . . straw-berries. There were wild blackberries and rosehips on the hill."

While John worked the mine, Maxine decided to build an art studio, for her painting and sewing and writing. "I had seen a bottle house in Rhyolite (near Death Valley, California) and that was my inspiration."

"I laid 2 × 6's on the ground, and 4 × 4's at the corners. Then 2 × 4's in between and I braced it, then started laying the bottles with mortar. Once I started building, people would stop by to see it.

Then they'd come back and bring me bottles."

John and Maxine would get snowed in periodically in winter. "Times do get tough in the west and winters, especially. There were times in the canyon when John had to walk out through three feet of snow to the top of the pass. Sometimes he could get a ride into town (20 miles) and a coupla times he walked all the way."

John had some pretty wild mining friends. "The Duffy boys, they lived in a cave, an old mine. One day they were digging and they broke through into a tunnel that led to another mine. They found out that it had been lived in by a man named Herman Napstein, known to locals as Herman the Hermit. He had gone back east with baby food jars full of gold dust, and was picked up by police in New Jersey as a vagrant. The entrance to the mine was all barricaded with rocks, and the only way they got in there was from another tunnel."

Bottle-Laying Tip from Ma

"10 Michelobs and 10 Millers do not a row make."
(A row of Michelobs = 8½", of Millers = 7¼".)

more...

A Letter from Maxine

". . . Everything John builds works so far. His Dad was out here first week of August, and shook his head sadly at all of John's good welding, "He could be making $75 an hour in the Bay Area welding." He didn't even remember his son was certified in the Navy for welder and pipefitter, and jungle fighter, also killer, thanks to Vietnam and U.S. Navy. I guess there are a lot of things we learn in a lifetime that our folks don't know. But we weren't happy living in the Bay Area, so now we live here. We cut wood all winter to keep warm, and I cook on the wood-stove, to save propane for JP's cutting torch. So what if we are different?? Pioneers never had it half as easy as we do. As I like to tell people "I've got my 30 years in with the PG&E." So I retired to a life of hauling water and chopping wood. I sometimes wonder if it would have been easier on John, if I let him put in his 30 years. His carpentry skills have got better the past 12 years out here, but nothing would pass a city inspector, I guess, but we are happy with what we make ourselves. Our plumbing didn't freeze and have to be dug up like most folks in town had. Our water runs year-round, freely. You could almost write a whole book on the things John's built from Fitting to our 23 × 14 cabin in South Fork, the bar and stereo cabin, to my book-house and the shop and front porch, to our bedroom, (best yet), to the present mill. Still building. Even better since we got your *Shelter* book . . ."

One of Maxine's dolls   *Pond above house*   *A psalm of Nevada*

◄ *Ma made this by cutting out aluminum foil, as with paper dolls, then gluing it to a piece of felt. (Turn it 90° to see the symmetrical pattern.)*

This old newspaper article was posted on the wall.

PROSPECTORS

Some people think that prospectors are strange, others say that we are just a bit off "upstairs." We are known for leaving the cities, the fancy clothes, barber shops, beauty salons, and MODERN CONVENIENCES like bathrooms. We drive from smooth cement roads to surfaces of rock and dirt and trails of dust — to pursue our dreams.

Lipstick and hair-dos yield to sloppy hats to shade our faces, (no longer made up) — the clean-shaven faces are exchanged for beards. Dresses and suits are traded for our shapeless chest waders and old dirty levis. Our once clean and manicured hands are now water shrunk and rough, we don't dare use 'lotion' as it might float away our gold flakes. Tables and chairs are absent, bedrock and boulders serve their purpose instead. Tree roots often hold our treasure with its 'colors' and black sand.

We hear music from the running creeks. We appreciate art as we look upon a hillside, and sculpture is seen in the crevices that we clean. When we look at a rock, it is not just to be stepped on, it has a story to tell with its smooth or rough surface.

You will recognize us with our shovels, pans, dredges or sluice boxes. Our dreams keep us going. Today, we have almost forgotten yesterday's sweat and toil. Sore backs and scratched knees, are but a memory. Today, the creek beckons as tho' to say, "Don't Give Up." We believe, on some gravel bar, a ledge, or hillside, maybe the next crevice or behind the next boulder, our gold will be found.

The Prospector is likely to be the most stubborn, closed mouth, but open eared and unattractive critter who walks the face of the earth. When we meet, we share an understanding look, a ready smile, a sparkle in the eye. If we are fortunate enough to compare our little gold bottles (even if they contain only a few colors) we know that we share a special friendship.

You will see us kneeling by the creek, for this is our usual position. It is also a position of being thankful, for we are truly grateful, and we acknowledge that we are, indeed, a guest of the land.

I'd rather be a prospector than prosperous in the city's. JP

We have done "So Much" With "So Little" For "So Long". now I can "Do Anything" with almost "Nothing" & Survive HERE

A 50-year-old outhouse

The bathtub in the creek was a great place to cool off on hot afternoons.

Bottle fence crowned with little cars

"I stay home and let the world come to me. Life's so good out here most times, I'm afraid to leave home in case I miss some new friend I haven't met yet."

I took my two boys up to visit the Pages one weekend. Evan, then about 8, followed John around wherever he went. (Evan always liked a "man's man.") Late one afternoon, they were down near the pond and the dog started barking. Turned out there was a rattlesnake in the bushes. John pulled out his pistol and shot it, and then that evening Ma skinned it and gave the skin to Evan.

Epilog: In the summer of '92, John Page died in a tragic accident. Maxine wrote, "He'll never be free to walk my valley again. He'll miss that and shooting rabbits and coyotes. But not as much as we will all miss JP." Upon John's death, the BLM cancelled their lease, and Maxine wrote: "I'm lonely and still wondering how I can go on without John. We had a hard life, but a good one and most of all, we had love . . ." Maxine moved to Henderson, a suburb outside Las Vegas. She's now doing OK and is building — yes — another bottle house, this time in her daughter's back yard, an "eight-sided gazebo." If you have any questions about bottle houses, or maverick mining in Nevada, you can contact her at: 130 E. Pacific Ave., Apt. 125, Henderson, NV 89015.

Jack Fulton and I went to John's wake in August, 1992. About 20 people showed up and we had a barbecue and hung out.

FLYING CONCRETE
Structural and Sculptural Forms in Lightweight Concrete by Steve Kornher

Outside view, my house. What will some day be the front door is presently bricked up forming a niche inside. Studio in background. Note the re-bar sticking out to add on the second floor later.

Former bedroom, now the living room. We keep adding rooms whenever the budget will allow — cash construction.

I ran across Steve Kornher's work on the web. Steve has been building for 30 years, 15 of those in Mexico. He's worked with adobe and rammed earth as well as various types of concrete masonry construction. He is now "completely in love with lightweight volcanic aggregate." Here is his account and photos of his latest work.

View of the picnic table from the roof at Tim's (about 3M diameter) This is ferro-cement, about 1½" thick and uses hard concrete with added concrete colors.

MY WIFE, Emilia, and I live on a two-acre *ranchito* about 25 minutes from San Miguel de Allende, in the mountains of central Mexico (6200 ft. elevation). Our home is a work in progress; built Mexican style: pay as you go and leave the re-bar sticking out for future additions. We built the house and large warehouse over a period of two years, using two (sometimes three) masons at a time. Building slowly is a lot more enjoyable and you can be more creative, since you can think things over and make changes.

The house presently has about 1200 square feet (110 square meters) of interior space with plenty of terraces for outdoor living. Most of the first floor is of adobe construction. Later additions and roofs are lightweight volcanic aggregate. South-facing windows and overhang provide passive solar heat in the wintertime. All roofs are masonry vaults with shell motifs. Mexico has some great masons and I owe a lot to the knowledgeable maestros who have helped me figure out how to do this wild and crazy stuff.

I originally came to San Miguel for a two-week ceramics workshop, got into potpourri and botanical exporting (all legal), then flower and seed production, and am now back at construction. I've been in the area 18 years and have lived at the ranchito for eight.

We have almost one hectare (two acres) here. About half is in flower production for my wife's store in San Miguel and the other half is largely native plants. It's a jungle during the rainy season. Between the flowers and natives, I'm shooting for 400 species.

* * *

One of my main goals is low-cost building construction that lasts 400 years. To accomplish that in this climate you need to build self-supporting structures and use masonry construction — adobe, lightweight concrete block, reinforced concrete columns, etc. The roof is the key to long building life, so it needs to be self-supporting, roundie-curvie — and not flat. Self-supporting vertical walls by nature want to be roundie-curvie. You go on from there and pretty soon everything is roundie-curvie. When you start to think about a long-lasting house, it's best to build so that remodeling is possible (probable). With adobe and/or lightweight concrete construction, you can hack out a doorway later on. With hard concrete this is an almost impossible project.

ROOFS

I have worked with a lot of different forming systems for different roofs. The largest to date is approx. 6M × 6M. Small roofs can be supported from above during construction, but larger roofs need some center support. Designing a smart roof shape is one key. Roofs in barrel vaults, modified domes, and especially seashell shapes (my favorites, and actually quite easy to do) are all very strong in compression. Since the roofs are self-supporting in shape and poured slowly, very little reinforcing is necessary. Once a lower form mold is made-up (typically pieces of ⅜" re-bar or welded wire), it can be easily moved in sections to form an identical roof.

Roofs are all built with an initial ⅜" shell poured on plaster lath on top of a metal framework (which is later removed and reused). This shell stiffens everything up for the pours of lightweight aggregate which follow and lets you see what the roof will look like. Changes are easy at this point. After the roof pour (4–6" thick) you can move the form again in five or six days for the next roof. I'm a big fan of reusable and movable formwork, usually ⅜" re-bar and/or #10 or #6 welded wire. These days I'm very excited about quick, low-cost barrel vaults.

WALLS

Walls are typically 6"–8" thick, built from the same volcanic aggregate. At first I poured the aggregate into forms (8/1 aggregate/cement by volume for walls, 5/1 for roofs) but now blocks are being made locally and since they are quicker, I use them for even roundie-curvie walls.

Books that have influenced my construction: *The Owner-Built Home*, Ken Kern: great alternative ideas about using concrete. *A Pattern Language*, Christopher Alexander. Any book on Antoni Gaudi; *Gaudi of Barcelona*, Rizzoli Int., N.Y. is a good one.

Studio door at our house. Iron, exposed lightweight concrete, fibers, and concrete colors.

The kitchen roof at my house. This was my first attempt at a seashell-shaped concrete roof. It ended up too flat and therefore needed a lot of iron in the reinforced beams in the ribs. This roof is built to support a second floor above.

Clay model of the first stage of construction of my house. I now use modeling clay.

Lightweight concrete roof over the closet area — three intersecting shell shapes. Note south-facing skylight.

Stairway built of lightweight concrete at Bonnie and Haden Kayden's home in San Miguel de Allende. We have since built a railing but it isn't absolutely necessary.

Window arches at Las Cañas. We nail up orange poly ducto to get an idea of desired shapes, then cut and fill with lightweight concrete.

Living room ceiling. This was the first brick bobeda roof my maestro or I had built. A challenging project.

Lightweight Concrete

Concrete is strong in compression; the best way to take advantage of this property is by building structures that are inherently self-supporting and don't need a lot of iron reinforcing. Since most building here in Mexico is with concrete, it is easier to let your imagination go wild. Local builders have been working with ferro-cement, wired styrofoam panels, plastered straw bales, and soil-crete. I have had the most success with lightweight concrete. Lightweight concrete differs from heavy concrete by its use of naturally lightweight materials (aggregates) such as pumice (volcanic stone) in place of the sand and gravel used in ordinary

structural concrete mixes. It weighs only half as much (50–80 lbs./cubic foot).

Not all concrete is ugly, hard, cold, and difficult to work with. There exists a whole range of lightweight concretes . . . "which have a density and compressive strength very similar to wood. They are easy to work with, can be nailed with ordinary nails, cut with a saw, drilled with woodworking tools, and easily repaired. We believe that ultra-light weight concrete is one of the most fundamental bulk building materials of the future." (A Pattern Language.)

Some form of suitable aggregate is available most everywhere in the world. Our locally available aggre-

gate here in San Miguel is a type of pumice/scoria (called espumilla or arenilla in local Spanish) which we typically mix 8:1 or 10:1 with cement for walls, and 5:1 for roofs. Most lightweight concrete has a good R-value and is a good insulator of heat and sound. It is used as soundproofing in subway stations. It has tremendous sculptural possibilities and is ideal for monolithic, wall-roof construction.

I feel that we need more intelligent building systems. I'm looking for a home that lasts several hundred years, that you can maintain and remodel easily, and that uses mostly locally available, abundant materials. Lightweight concrete fits the bill.

Styrofoam panels were assembled on the ground, then raised into place and plastered. This structure encloses a large, plastic water storage tank at Tim's.

In the foreground, the ¾" shell is in place. In the background, 3 inches of lightweight aggregate has been poured with one layer of #6 Welded Wire, then polished with a sand and cement mix.

Forming a modified dome at Las Cañas. Plaster lath on top of lightweight metal formwork (which is later reused). The initial ¾"-thick shell has been applied at the left of the photo.

The small kite/parachute roof for the laundry at our house. The only non-seashell roof so far. This is a ferro roof and the concentric circles of ⅜" re bar stay in place.

more...

View of the main house and privacy wall from the road. The neighbor's house in the right background is also built of lightweight concrete.

TIMOLANDIA

Photographs by Steve Kornher

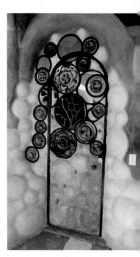

Tɪᴍ Sᴜʟʟɪᴠᴀɴ's homestead is on four acres, 20 minutes from San Miguel de Allende. He is Steve Kornher's neighbor. Present major constructions include the 70-square-meter main house, detached composting toilet, service room, carport, a small shop, and privacy walls. Nearby is the large shop with attached studio apartment.

Tim was interested in stretching the possibilities of lightweight concrete. He liked the curves of Steve's house and concrete doors and decided to go for it at his own site. It was an evolutionary, design-as-you-go–type project. Often models were made in the morning for work to be done in the afternoon.

Steve Kornher did the wall and roof shapes, working from Tim's "footprints." Tim and another neighbor, Robbie Friedman did the painting and a lot of the finish work. From the start Tim wanted bright, strong colors and where he couldn't get them with colored cement plaster he used various types of paint. All roofs were built with light-weight aggregate and the walls are either lightweight concrete or local rock.

The locals call the project "Timolandia." Steve wrote us, "It's been a lot of fun plus an incredible opportunity to experiment with different wall and roof forming systems here."

East side, main house

Ferro-concrete door. Lath and hard concrete.

136

Van Gogh's "Starry Night"? Well, almost. This is a sliding garage door; hammered metal, lightweight concrete, colored glass.

Stairs and rails, terrace at Tim's house. PV panels in the background.

Undulating Teja roof. Large shop at Tim's.

Stairs up from the kitchen at Tim's. These stairs are quite steep but the exposed ⅜" re-bar makes a good grabber without interfering with the tread. It works well, even for kids' feet.

Passageway between kitchen and living room. Shower is on left.

Above: "Cabbage Roof." Right: "Cabbage Roof" under construction, Tim's small shop. Lightweight aggregate made it easy to nail on heavy wire and plaster lath to form leaves.

more...

Stairway, main house. This started life as a "flying" stairway until it was decided to use the space underneath for a small gas heater enclosure, which is seldom used.

Snake gate and José, one of the maestro builders at Tim's

*Amazing curvilinear sculpture of concrete.
Saludos, bloqueros!*

Tim's kitchen

Abalone shell roof. Skylights, day and 12V stop lights, night.

*Structural beams support the lightweight
roof of Tim's kitchen, under construction,
below, and finished (at right).*

139

Cable bridge to little treehouse bar, with sleeping loft above, of "Big Treehouse in the Sky"

TROPICAL TREEHOUSES

David Greenberg

I**N THE '60s,** David Greenberg and his pal Roger Webster ran a company called Environmental Communications. They were set up in a loft in Venice, California and had a series of architectural slide shows they rented to schools and other groups. Included were slides from our book *Shelter*, as well as from Paolo Soleri, the Archigram group from England, the Ant Farm, and many others.

I lost touch with these guys for about 30 years until one day last summer when I came across Roger travelling in a painted-up schoolbus *(see p. 181 for the L.A. Filmmakers' bus)*. This led me to getting back in touch with David, who it turned out had built treehouses on his land in Hawaii, and was designing treehouses for an eco-resort in China. David has an architectural degree from Arizona State University, had practiced architecture in L.A. for several years, and taught architecture at UCLA for eight years. He gave up on all that seven years ago and moved to Hawaii. *(See the next page for the genesis of the treehouses shown here.)*

Four levels of "Big Treehouse in the Sky," eco-resort on Hainan Island, China. Rear view.

Treetops Treehouse, Hana, Maui, Hawaii

Hale Bar and Hotel treehouse in Maui, built by master builder Francis Sinenci, based on 2000-year-old Polynesian design

Front view, "Big Treehouse in the Sky," Hainan Island, China

Above and below: Three interiors of Hale treehouse in Hana, Maui, Hawaii

Another view of treehouse bar and loft shown in top middle photo

more…

In the mid-'70s I was picking magic mushrooms in a cow pasture on the island of Kauai when it began raining. I ran to a thick grove of trees for cover. I was soaking wet, but in the grove were a couple of guys that had already picked their limit of 'shrooms. When I told them I was in graduate school studying architecture at UCLA, they insisted I come and see the house they built; it wasn't far, they said, at the jungle's edge on a beautiful beach.

It turned out that the owner of the land had given about 50 hippies permission to live in the trees at the beach's edge, and they had built about 12 beautiful treehouses. I went from one to another in disbelief. (I was interested in "alternative architecture" at the time, but I hadn't seen anything like this.) They had used a lot of bamboo and clearish vinyl for the "roofs" to keep the rain out. We ended up in the nicest "house," theirs, climbing up a bamboo ladder to a "room" filled with throw pillows covered in Hawaiian patterned cloth and a grass mat floor with a pattern of leaf shadows made by the sun filtering through the trees above. I lay back on some cushions as they began to play some musical instruments, mesmerized by the kinetic shadows of the leaves moving on the floor, seemingly synchronized to the music. I'm sure the mushrooms had something to do with my total enjoyment of the environment but I felt I had finally found perfect architecture. After a while my hosts all decided to jump into the ocean across the little sand beach. I decided to stay a while and rest. After they left I began to hear the wind blowing through the trees above. It got louder as the shadows of leaves made moiré patterns on the walls and floor. The last thought I had before falling asleep was, what beauty. I woke to the sound of a big wave and went down to body surf in the warm water.

Five years later I bought a 20-acre piece of land in the jungle in Maui, and in 1996 I came across the picture I had taken of that treehouse. I was amazed at how beautiful it was. I had to find some bamboo. Fortunately there's a lot on this side of Maui, but the forests are

remote and hard to get to. Commercially it's $2/ lin. ft. I needed a lot of it, but had little money.

A few days later I got lucky. I had put the word out on the "coconut wireless"(the main form of communication in Hawaii) that I was building a treehouse and needed bamboo. It was winter and very windy that month. A friend had a friend that had a lot of tall, thick Golden Bamboo (the kind with intermittent green stripes) that was threatening the phone and power wires on his flower farm and he'd be happy if I carefully removed it. The next day I was there bright and early with a chain saw. The winds were gusting and blowing the bamboo against the wires. We had to tie ropes high up, then saw and pull just as the wind blew the bamboo in the right direction. We cut 50 stalks in two days, the average length 26′. We built a special wood frame on my Jeep Wrangler and hauled it back to my farm.

I began working feverishly, like I was on a mission, like a madman, getting up at the crack of dawn and working until it got dark. That next morning sunrise was oddly accompanied by the ring of the fax machine receiving a message. It was a poem about treehouses from a friend, a lawyer lady in L.A. The poems kept coming every few days and were the basis of a spiritual plan that made the treehouse feel so great to be in. Though the tree didn't become particularly more anthropomorphic, it did become more of an equal, a friend. Sometimes I would be up on the deck working and thinking about the next move or detail when one of the lines of a poem spoke the move and it was perfect and I would jump up and down with joy. The tree took the weight of the jumping. I felt like a kid jumping up and down on a bed for joy, and a little closer to heaven with each jump.

Every morning I was up before sunrise so that I could spend every minute of daylight working on the treehouse. As the site was literally in the middle of the jungle, it would get dark fast and I was always sad to stop working. I went to the dump every day.

Anything free was a potential building material. I also would spend time each week looking through dumpsters in the back alleys of the industrial part of Maui. I found 10 screen doors behind a screen shop, a major score, and the main sleeping loft became almost all screened in.

I have a nice hanging hammock chair with a great 200-degree ocean view below. At the end of one grayish day, I was in the hammock, rocking peacefully. It wasn't going to be a great sunset, but I was stuck with it. It was quiet, except for the sound of a few birds and the whisper of faraway waves. The bird songs gave me the idea to whistle. I started with tunes from the musical *South Pacific*. When I got to the tune "Bali Hai," I noticed a yellow-beaked mynah bird come fairly close to sit on a branch of a tree just a few feet away. I began to whistle to that bird, really getting into it. The bird seemed to respond to the tones, and jumped from one branch to another, coming ever closer. After a bit another bird joined in, jumping on the branch, then another. I ended up giving a concert of South Pacific to about five birds and didn't come inside till dark.

In a sense dealing with the treehouse was almost all details. Everything was customized. I discovered the best details by accident, in real time. There's a little "island" that protrudes slightly above the main deck because a big branch that the deck was resting on had a large gnarly elbow I didn't want to cut away, and so the tree expressed itself beautifully in yet another dimension.

Some of my friend's poems that were faxed during construction spoke of the parts of the tree in such loving terms, I now deified the tree as I did all of Nature. The nights I slept there were perhaps the most memorable of my life. As the moon filtered its way through the leaves and branches it seemed to light up the treehouse like an old-fashioned candlelit Christmas tree. And I was the present. When the sun would rise in the morning, it rose over the ocean and into the tree.   –David Greenberg

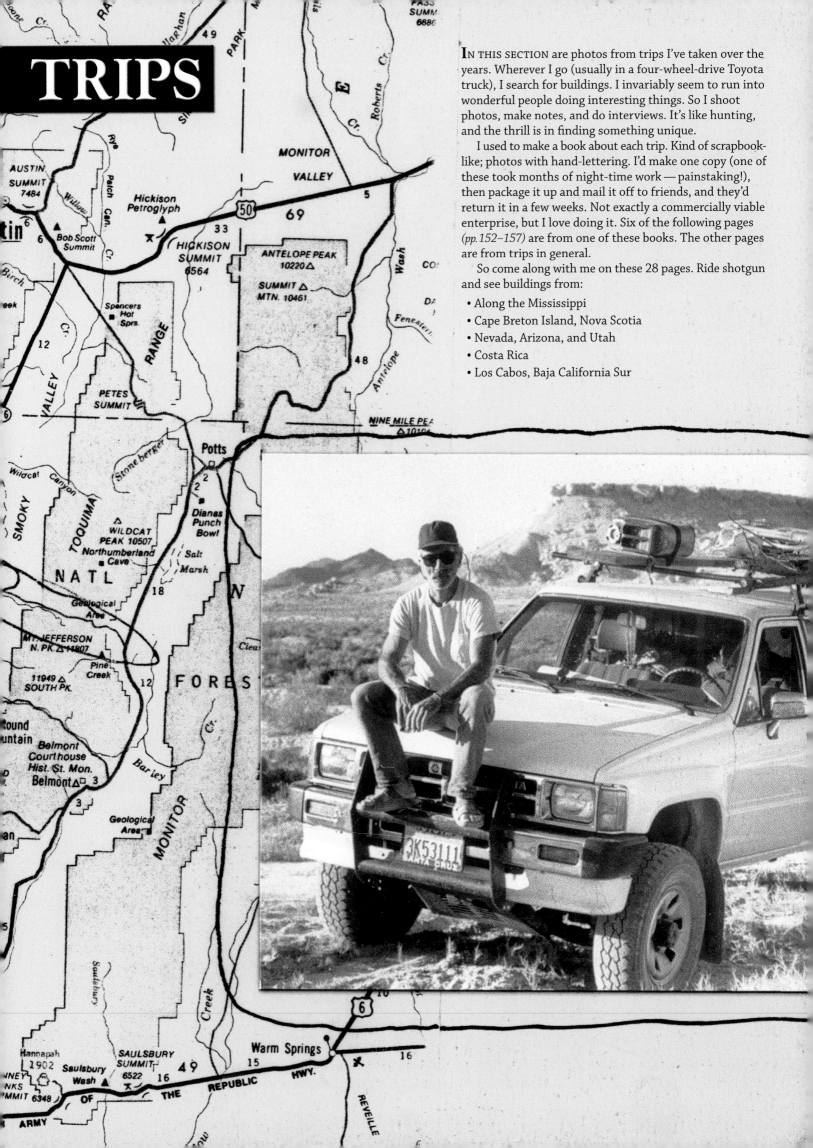

TRIPS

In this section are photos from trips I've taken over the years. Wherever I go (usually in a four-wheel-drive Toyota truck), I search for buildings. I invariably seem to run into wonderful people doing interesting things. So I shoot photos, make notes, and do interviews. It's like hunting, and the thrill is in finding something unique.

I used to make a book about each trip. Kind of scrapbook-like; photos with hand-lettering. I'd make one copy (one of these took months of night-time work — painstaking!), then package it up and mail it off to friends, and they'd return it in a few weeks. Not exactly a commercially viable enterprise, but I love doing it. Six of the following pages (*pp. 152–157*) are from one of these books. The other pages are from trips in general.

So come along with me on these 28 pages. Ride shotgun and see buildings from:

- Along the Mississippi
- Cape Breton Island, Nova Scotia
- Nevada, Arizona, and Utah
- Costa Rica
- Los Cabos, Baja California Sur

These buildings have a broad roof that acts as an umbrella, sheltering walls and windows from rain, and as a parasol shading the house on a hot day. The porches (galeries) were exterior hallways and cool places to sit on hot, humid evenings. Double French doors open in to each room; shutters open out.

ON THE RIVER

IN THE MID-'70s I went to New Orleans to do a slide show on our book *Shelter* at the Tulane School of Architecture. The day after the event, I rented a car and went out to explore the countryside. I drove up to Baton Rouge on the west side of the Mississippi. It was an overcast, humid day. Here, on the banks of this mighty river, things were strangely silent. There were very few people around.

Many of the houses, as you can see, were deserted. These were obviously homes of working people, and it was obvious that a way of life had ended. In the plantation house, there was a map on the wall from the 1800s, showing the division of land along the Mississippi delta: roughly pie-shaped pieces of land fanning back from the river. You could look at the map and picture the vitality of life along the river in those days.

In the field behind this house was the head of an old rusty hoe — a full 8" across. (A normal hoe is around 6".) It obviously took a very strong man to use a hoe like that.

How a slight bit of ornament can make a plain building special: here the circular design and rounded shingles in the eave, the bit of filigree woodwork at the top of the posts.

This large, well-built hip roof house was abandoned, and sat out in a field. Notice that there are no sags in all the horizontal lines (edge of roof, porch deck, railing). It was built way up off the ground. Mary Mix Foley, in her book The American House, *says that these raised houses date back to a very ancient peasant dwellings where the lower part of the house was a stable for animals and that in French America (as here), the lower area was often used for a laundry room, work rooms, or storage.*

Small, tuned-in house in the same neighborhood as the mansion shown below. Note similarities: roof shape, dormer windows, columns, porch area — and the symmetry.

Oak Alley Plantation, Vacherie, Louisiana. Built 1836.

A nice broad galerie, with arches made by ornamental woodwork

145

NOVA SCOTIA

This was on the French section of the island. The main roof is a high gable shape, but with the spectacular expanded dormer window (très élégant!), and covered porches. Everything is perfectly symmetrical: chimneys, deck posts, windows, deck railings. The central porch roof is designed so it appears to flow out from the bottom of the dormer. Rocks on the grass are painted white.

In SUMMER OF '73 I worked with architect Bob Easton to produce the book *Shelter*. Then next summer I worked for Stewart Brand as an editor on the *Whole Earth Epilog*. Stewart had just bought some land in Nova Scotia, on the west side of Cape Breton Island, facing Prince Edward Island across the Gulf of St. Lawrence, and I agreed to meet him back there at the end of summer and help him build the foundation for a (Bob Easton–designed) house.

My son Peter, then 12, and I took off for points east on the wonderful trans-Canadian train. We got off the train in a small town north of Toronto because the train conductor told me about some great barns in the area. We spent a few days shooting barns, a few days in Toronto, and then found a 45-foot-long school bus that needed delivering to Nova Scotia, and off we went. Stewart's land was on the west side of Cape

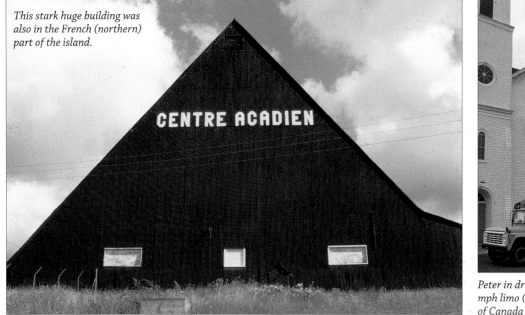

This stark huge building was also in the French (northern) part of the island.

CENTRE ACADIEN

Peter in driver's seat of our 45'-long, 50-mph limo (a governor) across 1000 miles of Canada

A much simpler version of the house on the preceding page. Again, dormer window is main element added to basic gable shape, and the dormer window roofline curves out slightly at the bottom and gives it some flair. In all four designs shown on these two pages, the dormer is brought up flush from the front wall. This is a nicely designed little home.

Breton Island. He had a rattletrap Ford van with holes in the floor in storage and we fired it up and drove around exploring the island. The quality of building — homes, barns, outbuildings — was amazing. A tough climate, scary severe winters, fierce Gulf storms, short growing seasons — no room for mistakes.

Scottish, Irish, and French people had migrated to the island long ago and many of their descendants are still there. Photos of settlers around the turn of the century show what looks like a different species of humans: giant, strapping men and tough, resilient women. They cleared forests, planted crops, built homes and barns, caught fish and lobsters, made just about everything for themselves, survived the winters. The architecture on the following pages is obviously a legacy of these strong people.

Three dormers and a lot more ornamentation, including black-painted highlights and gingerbread woodwork. Lathe-turned pieces at dormer roof ends carry through the scrollwork theme to give it a lace-like effect.

The same basic design as the house on the opposite page, but simpler. Stumbling across buildings like this is like finding a treasure chest. (Dormer windows flush with exterior wall seem characteristic of the area.)

more...

A soulful Cape Breton barn, in tune with its surroundings, elegant in its simplicity. Almost a saltbox shape, but the roof breaks angle about 10' down from the ridge on the long side. Square window placed point up at eave is fairly typical of region.

Pure Architecture

Stumbling upon an area with buildings like this is like discovering a treasure chest. When I went to Cape Breton Island, I had no idea that I'd run across such wonderful farmer/builder architecture. These buildings not only look good, but they're built well and are functional. Moreover, they are instructive for home builders. Many building shapes you see in the countryside could be models — with adaptations for solar heating, insulation, etc. — for building a home in the same area.

Isn't it strange we never see buildings like this in books or magazines? And also strange that so few architects seem to know the definition of architecture: the art and science of *building*.

This beautifully proportioned and detailed barn would make a nice house shape. The gambrel shape (where the roof changes pitch), gives you more headroom for hay (or bedroom space) on the second story. The dormer is simple and straightforward: an extension of the upper roof line and the front wall line. Note slight upturn at roof's edge to shoot rainwater out from walls.

A nice little gambrel-shaped outbuilding, with red-stained walls and white trim. The edges of the roof (and the barn at left) are finished with fascia and soffit boards so there are no exposed rafters or roof sheathing boards. This gives these buildings a tight, clean look.

I love this little building! The proportions, the simplicity, the way it stands in the field. Unselfconscious, perfect architecture.

A lobster fisherman's storehouse, symmetrical hip roof

Another tight gambrel shape. They call the slight upturn at roof's edge "flying gutters." (Notice this detail also in the two buildings on bottom of the opposite page.) The builders did this by nailing short two-by-fours at angles at the rafter ends. It shoots (most) rainwater away from the walls and windows, and provides a nice visual touch.

more…

Above and below: Local color

An abandoned church, nicely designed and executed. It was built, like many of the older buildings on the island, on a foundation of huge sandstone rocks, laid flat on the ground.

This guy had made his mailbox in the exact image of his house and placed it on a post symmetrically in front of the house. Here he's standing on the front steps with his three kids.

A charming little symmetrical hip-roofed log cabin

George built this decorated little cabin for his mom in New Brunswick.

George and Mom

A silver-painted bus converted into a home

Next to the log cabin (at left) lived Frank, a retired machinist, who built his home out of local (spruce) 2 x 4's and pieces of sheet metal for $35. He used a wood cookstove and had an abundant supply of firewood tucked under and around his silver house.

These six pages (152–157) are from a hand-lettered book I made of a trip to Nevada, Arizona, and Utah in 1989.

Barn with soul, Escalante, Utah.

It seems to take about a week to get into nomadic gear. By then you're a lot more tuned into the desert, the roads, the people and the ways & daily routines of western travel.

Bryce Canyon is a scenic wonderland all right, but again the heavy hands of the park service bummed me out. I mean, do we really need a "News of the park" FM radio station? Decided there to settle for less spectacular but also (therefore) less commercialized terrain.

Escalante is a great little Utah town on Hwy. 12 between Bryce & Capitol Reef National Park. It's near the turnoff to the Burr Trail, a spectacular 70 mi dirt road to Capitol Reef.

Saturday morning, 9:00, sun is shining, am heading up dirt road along Cottonwood Creek toward Henrieville & Bryce Canyon. Bob Dylan & the Wilburys are singing "Tweeter & the Monkey Man" & I reflect on the moment and my good fortune. A 4x4 THAT can go practically anywhere, food, water, tent not to speak of radio, tape player & 3 Coronas in ice chest. I mean...

Inside above barn

Roamed around Escalante shooting pics, then got groceries & headed out of town to camp.

Elegant

Built in late 1800's, according to a neighbor. A crisp building--look how sharp all structural lines are--no sag.

ESCALANTE,

Escalante is one of those towns that has it. You come across them occasionally in travels. They're special. They feel good. It's the way they're sited, the buildings, the people, the town's history. And people that live in these towns invariably know what they've got.

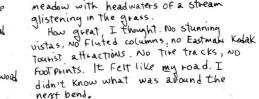

Part way up the main Escalante canyon was this old Homestead about ¼ mi. off the dirt road. Drove across the field & 4-wheeled over little creek, then drove up an old logging road a mile or so. Set up camp with great feeling of freedom. No one within miles, in fact it looked like no one had been by this way for years.

Built fire with abundant cedar deadwood lying about. What a smell! Cooked dinner, rolled out sleeping bag, with help from a star gazing book, saw Ursa Major (big dipper) for 1st time in its entirety.

(THERE'S A LOT MORE TO THE CONSTELLATION THAN THE DIPPER PART.)

Next morning walked up the logging road a few miles to a mountain meadow with headwaters of a stream glistening in the grass.

How great, I thought. No stunning vistas, no fluted columns, no Eastman Kodak tourist attractions. No tire tracks, no footprints. It felt like *my* road. I didn't know what was around the next bend.

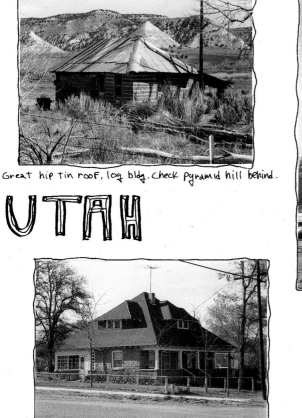

Great hip tin roof, log bldg. Check pyramid hill behind.

UTAH

All right!

In fact, a town of hip roofs, plain & fancy, as above. This is an unusually large equilateral hip roof. Dormers repeat roof shape nicely. Nice porch in front, glassed-in porch in rear. Note white pillars front & back.

TORREY, UTAH

Slept out in a field off Hwy ⑫ between Boulder & Torrey, Utah. 8500 ft., coldest night of trip -- about 30°. Got up in morning & got into Torrey around 7:30 A.M., approaching town through valley with green fields, well watered.

Another beautiful town, wonderful feeling. Three-foot-wide ditches running alongside sidewalks, filled with clear, fast-flowing water. My mom, born in salt lake city, used to tell us kids about "water running down the streets" & I always thought she meant rain water in the gutters. But no, the mormons just diverted water, where it was abundant, & ran it throughout town -- brilliant!

Down a side street was this stop-you-in-your-tracks beauty of a house. This is one of those buildings I run across every once in a while in photo travels. The place fairly glowed. Does it show in the picture?

The log work is exquisite -- competent, quick, home-made & with tight joints. Check out notched-in log wall in middle (it's an interior partition, and provides tie-together bracing. what I really love here are the notched-in log cross-ties (count up to the 13th log on side wall). Somehow, that frosty morning, you felt all the lives that had passed the years in this little place.

Here, in one of the upper, drier parts of Capitol Reef, was this beauty. It was designed & built so well that it looked like it grew there. More good Mormon craftsmanship. The lines of the building are still straight & true. No sagging roof, no crumbling walls.

Curved framing →
for slight, elegant
upturn of eaves.
for one thing, this
shoots roof-caught
rainwater out &
away from walls.

Double-angled
notches

Ruby Santell

Across the road was this finely-crafted
little log church. Every part of it beautifully
done— windows, sills, door framing, foundation stones
graceful curve of eaves. Ruby Santell, raised-but-
not-born in Torrey said the building was now
owned by the Daughters of Utah Pioneers, that it was
used for dances & quilting bees until 1980. "Now the
kids just want to get away from it all," she
said. "Dogs, radios, dune buggies..."

Now here's some architecture!
Sited perfectly in canyon
Lightly framed upper story
Floats above house.

— on side road in Prescott Valley, Arizona

more...

At left is Kevin Hickey, who lives with wife & 2 kids in the house. He rebuilt the arches, cutting blocks ▭ -shaped with diamond saw & chinked walls with mud. Out back is his VW repair shop, called "Marginal Motors." He runs the hot springs water thru pipes in floor of shop for radiant heating & thru radiators in house.

He told me about some mining roads southeast of there with huge stone embankments built by chinese. "No one's been on those roads for 50 years."

About 4 mi. west of Benton is the tiny town of Benton Hot Springs. The creek at left flows with scalding hot water. The little house is built of lightweight tufa block & has in past been a bank, whorehouse & butcher shop.

General store/gas station

memories...

The expandable house -- in a meadow near Mono Lake ↑

Back Alley, small town, USA

A couple of ladies who collect junk run this place -- and have fun!

Some 2000 acres, including Benton Hot Springs, are owned by the Bramlett Family, Kevin's landlords. There is a big hot springs pond, but it's off limits to tourists now.

Old wagons outside general store. Except for road sign, this shot looks like a painting.

Nice old buildings in town. From here I headed west towards Bishop-- The Sierras felt like home.

Bishop, California-- some old fashioned carpentry.

Kurt and Shirley Van Dyke sitting on the second-story porch of Kurt's Puerto Viejo Hotel, about ¼-mile from Salsa Brava, a world-class, red-hot surf spot. Inset: Kurt, Shirley, and friends. When I first got into town, and was standing looking at the place, Kurt came out on the balcony and said, "Classic, eh?"

Johnnie's Discotheque and Chinese seafood restaurant in Puerto Viejo. This great building right on the shores of the Caribbean also contained the town general store (groceries, hardware, clothing) and a pool hall. It was owned by Manuel Leon, Chinese, born in Puerto Viejo 50 years ago. You could buy a beer and sit on a bench on the porch looking out at the aquamarine Caribbean . . .

COSTA RICA

In spring 1991, I took a surfboard and backpack and flew to Costa Rica. I intended to go to both the Pacific and Caribbean sides of the country, but when I got out to the Caribbean, with its beautiful waves, exotic jungle, and laid-back atmosphere, I ended up staying there the entire three weeks.

A bar playing reggae music overlooking a black sand beach with green waves breaking on the shore . . . going up a clear and sparkling river between Costa Rica and Panama in hollowed-out log canoes . . . walking for two days on a jungle trail into Panama . . . eating meals in tin shack restaurants on the beach with cool breezes blowing through the open walls . . .

I went out to the small town of Puerto Viejo to visit my friend Kurt Van Dyke, who had opened a surfers' hotel. I met Bill and Barb Castle *(see pp. 16–21)*, who were running a tropical bed and breakfast and taking impromptu groups of visitors on river canoe trips. I hung out for several days in Puerto Limón, an old port town with soul. I travelled around in a rental car and shot pictures of houses, which were mainly up off the ground both to avoid termites and to catch the ocean breezes.

"They went up this huge tree. The 'shooter' went up about 70 feet with a 10" slingshot and some ammo (rocks). One kid went up about halfway and the third stayed on the ground.

"He fired all his ammo at the iguana and hit it once. Then he dropped a handkerchief to the ground man, who filled it with ammo and then threw to the middle man. He then ran it up to the guy at the top.

"They were so far up I couldn't see them, and it was getting dark. He shot rocks and I heard 'thunk,' and then a crashing, and I thought 'Oh, God, the kid fell,' but it was this huge iguana. We cooked them in the hotel that night. They were great!"

–Kurt Van Dyke

Puerto Limón

Kids told me this was the house of El Gordo.

Tranquilo

Sunny/rainy day, having a beer in a little open-air bar in Punta Mona (near Panama) after a hike through the jungle. A bare-chested working guy carrying a machete comes in, sits down, looks at me and says *"Como está?"* "Bien," I reply, *"y tú?"* *"Tranquilo,"* he says.

What a wonderful reply. You ask a guy how he is, and he replies, "I'm tranquil."

Bill Castle's "Roll Your Own Jungle Tour": 13 of us went in one of these 35' long canoes up river from Sixaola, on the Costa Rica–Panama border. Shown also are Bill (with parasol) and Victor, our skipper. These graceful canoes are each made out of one hollowed-out log and powered by a 25 hp. Evinrude. We sat on sticks wedged between the sides. (See pp. 16–21 for Bill's log home in the U.S.)

House in the jungle between Manzanillo and the Panama border

Octagonal house north of San José

Escuéla Metallica, San José. Pre-fab metal building designed in France (by Gustav Eiffel, according to one account). Built in Colombia in 1890 and parts shipped to San José where it was erected. When you look in the windows, there are children in each classroom and a sunny courtyard in the center. It's in beautiful condition. A remarkable building.

Nice hip roof, tile shingles, open walls for breezes

Built to catch breezes, but top story looks under-supported.

DEEP IN THE HEART OF BAJA

In 1988 I bought a Toyota four-wheel-drive pickup truck, added a camper shell, and that spring headed south for Mexico. My destination: the southernmost tip of Baja California, called Los Cabos. I drove down Highway 15 on the mainland and caught the ferry across the Gulf at Guaymas. From the moment I got out on the blue waters of the sea I was hooked. I got off the ferry in La Paz and headed south. Around Todos Santos, a Spanish-style town of elegant old thick-walled adobe buildings, you cross the Tropic of Cancer, and the landscape becomes "tropical desert," with a profusion of exotic plants, many of which burst into vivid bloom with any rainfall.

I camped on the beaches and in desert arroyos, went surfing and swimming in the warm water and pretty much liked everything — the land, the people, the water, the desert, the dazzling blue daytime sky and the clear black night sky with shimmering stars. Being a native coastal Californian, it felt kind of like home turf, but warmer, drier, more exotic.

One afternoon I went into a gift shop in San José del Cabo and started talking to the proprietor. Isidro (Chilón) Amora Aguilar, about 30, had come to Baja from his native Mexico City in the early '80s, sold fruit on the streets, ran a restaurant, and now gift shop. We were both interested in what Chilón called "the real Baja," which is off the tourist track. He was going out to rock art sites, knew about remote villages, water-filled arroyos, working ranchos, fossil areas. I had found a soul mate.

Over the course of about 12 years, Chilón and I criss-crossed Los Cabos. We found cave paintings, went up water-filled arroyos deep in the desert, visited ranchers, covered thousands of miles of remote dirt roads, camped out, got lost, had cars break down . . . .

The Posada Señor Mañana, a "five-coconut hotel," in San José del Cabo, a bohemian hotel frequented by surfers, fishermen, and world-travelers. At right is Yuca (Rogelio López Rodriguez), proprietor/wheeler-dealer. At left is a cyclist who had flown from his home in Reno, Nevada with his bike, and was doing a circumnambulation of the Los Cabos area. For six years, I rented an upstairs palapa room with palm frond roofing and open walls for $1000 a year.

In 1983, Chilón got his own radio show, for young kids, on the main Los Cabos station. He assumed the persona of a parrot and called himself *Periquín*. He played good music for the kids, who would call in and talk to the parrot; the program ran for two hours each Sunday morning and was an instant hit; kids out on isolated ranches loved it. Subsequently Periquín became a minor celebrity, everyone knew him, and he knew everyone and what was going on all over Los Cabos.

Since I could fly down in 2½ hours from San Francisco, the idea was to leave a vehicle down there. My first car was a "Baja bug," a VW bug with fiberglass fenders and hood, big tires, and

Yuca, left, with Chilón on the radio in 1993

Desert landscape sunrise on the road to Bajia de los Angeles, with boojum trees — somehow perfect this morning with empanadas (small fruit-filled pastries) and tequila

The small town of San Antonio, off the beaten path near El Triunfo, a real Baja town unchanged by tourism or outsiders. Church on left, gas pump, town plaza, Spanish-style buildings typical of Los Cabos.

Primitive rock shrine just north of Santa Rosalía on barren hillside

Lady at her remote ranch on the Naranjas road, which runs from San José del Cabo across the Sierra de Laguna mountain range to the Pacific coast near Todos Santos. The ranch house and fence were built mainly with palo de arco branches, the house roofed with palm fronds. Almost all the ranch houses in Baja were built with materials that grow on the land surrounding the houses.

huge shocks. I had a rocket box luggage carrier on the roof wirh a marine photovoltaic solar panel that charged two heavy-duty batteries. It was a great little car but unfortunately ended up underwater in a huge unexpected rainfall/flood, so my next vehicle was an '83 Toyota 4×4 that I would leave either at Yuca's hotel or Chilón's place; I'd fly in, unwrap it, and take off camping.

I found myself getting drawn back to Baja time and again. Whenever I'd get a week or two clear I'd be off. Usually I'd fly but about once a year I'd drive, stopping at hot springs south of Ensenada, at Mama Espinosa's restaurant in El Rosario, at the beautiful oasis town of San Ignacio to sit in the shady town square. I'd drive out remote roads at night to sleep under the stars, go surfing, visit old missions . . . until I finally hit my southern destination, Los Cabos. On the following pages are some images from these travels.

www.shelterpub.com/_baja/baja.html

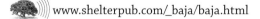

Typical Baja!

Peyton Carling standing outside the house he built in La Purisima, a beautiful oasis town in a desert arroyo north of Cíudad Constitución. Peyton raises organic oranges and avocados and runs eco-tours in the area.

Nicely designed hip roof house with upstairs bedroom, thatched roof, north of Todos Santos

Buildings of Baja

Below: Sensible shelter for absentee gringo. Lockable concrete block storeroom with heavy door, attached palm frond shaded area. Upon arrival, get out hammocks and barbecue and enjoy outdoor living in the desert.

House outside Mulege

Allan and Jeannie Maxey's palapa at Shipwreck Beach, east of San José del Cabo. Shelter in much of Baja consists of a roof, with no walls. Rafters here are red palm, purlins are carrizo (bamboo) tied on with hemp rope; palm fronds tied to carrizo with datillo leaf. This is traditional roofing technique for the area.

164

Floor tiles and bricks ready for kiln, San José del Cabo

Richard and Rae's palapa at Shipwreck Beach. Construction is basically same as palapa at the bottom of the adjacent page. This building is sited on the crest of a hill looking down at the ocean (and a good surf break) and it gets a nice breeze through the open walls on hot summer days.

Artist Alfredo Ruiz built this round palapa north of Todos Santos as an art gallery. The rafters are palma albánico, the purlins are palo de arco (sturdier than the usual bamboo).

Cobijos de Baja

Construction of the El Paraje restaurant in San José del Cabo in the late '90s. This is typical southern Baja ranch construction, called chiname: palm post and beam framing, with palo de arco branches interwoven as backing, then plastered with mud. The same basic technique is called "wattle and daub" in England. Roof consists of palm rafters, carrizo (bamboo) purlins, thatched with palm fronds. All materials are from the desert.

Corner of building showing corner post, woven branches before plastering

Adobe stove

Main house, adobe walls
and palm frond roof

Los Ranchos
de los Cabos

WHEREVER there is water in the Baja desert there is a rancho. You run across the ranchos in very remote spots, often unexpectedly. They typically run cattle that graze on desert vegetation, but ranchos in places with adequate water and soil grow vegetables and fruit. Some of them have goat herds, and make delicious cheese. The ranchos are completely tuned into the desert environment, the houses, fences, and many implements built out of local materials. It's a way of life going back hundreds of years, starting when the Spanish settled in Baja in the 1600s, and not too much has changed since then.

Rancho south of Mulege

Rancho on the East Cape of Los Cabos. Notice how neat and tidy and integrated with its surroundings it is. After a while you begin to see the beauty of the Baja ranchos, even if it's not the usual vision of European or American ranches with green fields. The buildings and grounds and fences are in harmony with the desert and its plant life.

Rancho Vinateria, a beautiful and productive ranch north of Cabo San Lucas. In the shaded building compound it's peaceful and cool in the hot weather. Photos of main house and kitchen below are same ranch.

Kitchen

Rancho near San Luis Gonzaga, with palm-thatched veranda connecting two buildings

Goat corral north of San Luis Gonzaga

En la Playa
On the Beach

One afternoon I started to set up a campsite at this beach ▲ at San Gregorio, south of San Juaníco, but a hurricane (chubasco) was forming out at sea. When the umbrella went airborne, it was time to leave, so I packed up and headed south. As the rain started to fall, it unlocked fragrant smells in the desert. By the time I got to the safety of a hotel in Puerto San Carlos, the wind was tearing palm fronds off trees and sending people running for shelter.

Steve Warren runs the Magbay Tours (www.magbaytours.com) surf camp on Isla Magdalena, an island west of Puerto San Carlos. There are usually perfect waves breaking at this remote site and Steve takes a maximum of 10 surfers out for a week, supplying food, shelter, and beer. There's practically no other way to get to the place, so surfers have the waves to themselves. Steve is a rare type of American, married to a Mexican woman, fluent in Spanish, well-liked by locals.

Pacific Boxfish has bony exoskeleton.

IT TOOK ME just a little over two hours to fly to Baja from San Francisco. I would leave home on a cold and/or wet day and a few hours later step out into the hot sun and cobalt blue skies. Chilón would pick me up at the airport; I'd take the tarp off my vehicle and head for a campsite on a remote beach. I carried two surfboards and would set up a flea market sun tarp and spend several days alone on the beach, surfing, swimming, beachcombing, enjoying the warm water and pristine beaches, the desert solitude.

I found the west coast of the Baja peninsula generally more interesting than the gentle waters of the Gulf on the other side, and I roamed up and down the coast over the years, often four-wheeling it to get to otherwise-inaccessible camping spots.

Sunrise on my beach

Camping in style! This guy was a former Greyhound bus mechanic who had bought the bus for $40,000 and he and his wife were camping throughout Baja. Greyhound busses of this era have air suspension systems for a smooth ride, and the bus was immaculately fitted out, neat as a pin inside.

Surfer's almost-invisible summertime semi-permanent camp setup in an arroyo next to a good surf spot, nicely tucked under shady trees

My second vehicle, this '83 Toyota 4-wheel-drive truck: the perfect Baja conveyance. The tent on the roof was made in Italy (see p. 180) and compacts neatly for travel; a great place to sleep up off the ground, catching breezes. The 12' × 14' flea market sunshade with 1" electical conduit poles and an aluminized tarp held on by ball bungees (see p. 180) is a cheap portable camping shelter for hot climates. Each of the corner posts is held down by a hanging canvas bag filled with sand. It all folds up and fits into the rocket box luggage carrier. Tent opening faces surf break.

A beautifully sited Mexican fishermen's shelter, built out of driftwood, and on a remote beach south of Punta Conejo. If this place weren't so hard to get to, it would be a million-dollar gringo building site.

My first vehicle in Baja was this little white Volkswagen "Baja bug." The rocket box on top held camping equipment and had a solar panel that charged two heavy-duty batteries. It had been built for scouting the Baja road races. It had a 15-gallon gas tank behind the rear seat, and huge shocks that came up and tied into a roll bar inside the cab. It was a great Baja vehicle until it ended up underwater in a flood.

Master bedroom in above fishermen's shack

Fino and Cleo Green. Fino is a 5th-generation local guy (descended from a British whaler), a tuned-in-to-the-area surfer, diver, fisherman, and owner of the Killer Hook surf shop. We'd go on camping trips to little-known surf spots. (For interview with Fino see: www.shelterpub.com/_baja/fino_josefino.html).

Zelate tree sculpted by wind en la Costa de Los Cabos

more...

At left is rock reef and beach where fishermen launch their boats.

Paradise on a Shoestring

WHEN YOU DRIVE to Baja you generally enter at Tijuana and drive down the coast to Ensenada. Much of the beachfront property is in American hands and covered with unimaginative houses. However for several years I had noticed a beautiful point with colorful ramshackle buildings, looking totally different from any other coastal structures. On my last trip I stopped by for a closer look. There were houses, restaurants, little stores, a fish camp — all built out of scraps of plywood and corrugated tin and painted with bright colors.

It turned out that a group of 50 Mexican families had formed a company and bought the land in the early '80s and built their homes there. It was a beautiful little community, built with very little money, and unusual in an area otherwise populated with condos, time shares, and tourists.

I had a cup of coffee and talked to the owners shown in the photos here about their unique community. They ran the place by monthly meetings and each family was allowed one representative. Americans

Sandy beach on other side of point

were welcome to visit, they told me, and to eat at the restaurants, but the owners were there to stay — the land was not for sale, at any price.

It's both unusual and heartening to see locals seizing a piece of their own land like this in Baja. Just about all the coast for miles north and south of them is owned by Americans, much of it fenced to block anyone from using the beach. Here you have a little working community, with fishermen, cooks, mechanics, storekeepers occupying a site where Club Med would die to put a hotel. *Si se puede!*

> *It is impossible to account for the charm of this country or its fascination, but those who are familiar with the land of Baja California are either afraid of it or they love it, and if they love it, they are brought back by an irresistible fascination time and time again.*
> —Earle Stanley Gardner

Looking back at land from reef

Heading up a gravel road south of Eureka bound for Elko via Jiggs

Jack

Jack's Mitsubishi pickup

COWBOY POETRY FESTIVAL

Jack Fulton and I have been taking occasional photo-shooting trips since 1972, when we took a ten-day trip to New Mexico to shoot photos for *Shelter*. *(See the last page of* Shelter *for the buddha-wanderer-wiseman-hobo we met on the way home.)* We have a great time on the road. Shooting photos gives us a purpose and invariably leads us to meeting people and seeing things we'd otherwise miss. Jack is a fabulous photographer, painting pictures with his Pentax, and it sharpens my eye to travel with him.

The last trip we took was in January, 2002, to the Cowboy Poetry Festival in Elko, Nevada. We left just after a huge snowstorm, and Nevada was spectacular with its deep snow and blue skies (and –50° wind-chill-factor temperatures). The poetry festival was unique. Cowboy poetry, continuous country music, the elegant cowboy style of dress of the Elko buckaroos, friendly locals. We shot barns in the snow and little houses in the towns, signs and gas stations, and cows in the bright morning sunlight. On this page are a few pictures from the trip. *(On the Road with Jack: A PhotoJournal is due out around 2010!)*

Left: Basque sheepherder's wagon, pulled around with the sheep by a horse; a spiffy little rolling home

I'll admit to a Baghdad Café fantasy, running a restaurant/bar/gas station/motel/store out in the middle of the desert. An oasis. Here one is ready to buy and fix up, in Mesquite, Nevada.

ON THE ROAD

DONKEY TRAIN ACROSS AMERICA

The Odyssey of John Stiles
Photos by Janet Holden Ramos

IN THE LATE '80s I heard about a guy travelling across California in covered wagons pulled by donkeys. This I had to see, so I drove up to Santa Rosa and found him by the side of a country road. John Styles, then 42, and a native of the Ozarks (one of 16 children), had been on the road for about 10 years, and had covered over 10,000 miles.

Stiles had 14 donkeys, 3 mules, 34 chickens, 3 goats, and 9 doves ("for music"). He built his steel-wheeled covered wagons on restored turn-of-the-century frames. He had spent time living with the Amish people in Illinois, learning the arts of self-sufficiency, and there was a definite Amish cast to him, including the beard and the self-discipline.

When he finished his morning chores, we sat in one of the wagons (it was cozy and had shelves lined with books), and talked.

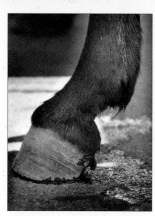

Goat leans out to nuzzle John as he tends to one of the donkeys.

LLOYD: *Do you always walk?*

JOHN: Never rode a mile in the wagon. I walk behind the mules when I've got the wagons rolling, both to save the animals from the load and to get that special perception you get when you move on foot.

You see I get up about the same time everybody else does to go to work. Five o'clock and if it's still dark, I light a candle, read from the good book and prepare myself for the day's journey. Then when it starts getting daylight I bring the animals in and the first five hours I work here in the camp. Then I get out and walk on that pavement with this entourage down the road four or five hours, two miles an hour, eight or ten miles a day and then I take another four or five hours, break it down, take it apart, and put it to rest.

People don't believe I live this way. It was a matter of conscience, I wanted to be self-sufficient.

I saw the world and how plastic it was and I wanted to stay true to my convictions of being close to the earth. I travel on the freeways and in huge metropolitan areas with no driver's license, no registration, no insurance, no taxes, no permits —nothing. Just totally free, and that's in a world where electronic mankind has begun living in a global concentration camp.

The question is, how do we transcend that? Maybe we need a return to some solid reference points. Maybe that's what I'm doing now.

What were you doing in between 1965–70?
I was on Haight Street all through 1967. I met real flower children, not hippies, but the ones who wanted to find themselves, be creative and who developed the "back to the land" movement.

When I left Haight-Ashbury, I went out into the country and became totally self-sufficient and, to me, self-sufficiency is not a four-wheel-drive truck and a chainsaw and a Troy-built rototiller. To me I had to use animal traction, I had to raise enough feed to not only feed the animals and myself all through the winter but to have seed left over to plant again and enough to eat until the harvest came again. I wasn't being honest with myself — I always liked playing around and this whole urban generation that went out there in the mountains and stuff, half-assed, no scams, no welfare, none of that stuff . . .

What do you do about food?
I never ask for food, I never ask for clothing, I never ask for anything in the world. People just come — like that guy this morning — he just dropped off a bale of hay for the mules.

I haven't spent one single penny on animal or people food all summer. I have eggs from the chickens. I drink a half-gallon of goat milk a day. I mean that is good. That's the key, that's the way out. It's the most complete nutrition known to man.

You're not going to do this forever, are you, like for another 10 years?
I'd like to settle down if I can make enough to get a little piece of land somewhere in New Mexico, put an adobe house on it, put a up windmill, build a tower and have water gravity. Have my little alfalfa patch, my goats, an orchard. Just a little tiny five or seven acres, you know . . .

Last time I talked to John, he was working on a book titled: **Are You Just Travelling or Are You Going Somewhere?**

Epilog: Seven years later, John had farmed for two years, found it too tough, and was back on the road. Here's part of an article about him that appeared in the *La Jicarita* (NM) *News* in 1996:

". . . he espouses his own particular brand of philosophy, backed up with extensive readings of both European and American writers who have warned about the dangers of society's increasing dependence upon technology. He has no use for computers, of course: He believes that the 'global electronic concentration camp' that they are creating will make decisions affecting all of mankind; about the economy, education, transportation, nuclear defense, medicine — without any moral or ethical criteria. 'Nobody is questioning if we should be doing all this. The only question being asked is can we.' And without your 'bar-code microchip laser-beam tattooed implant and your holy trinity of personal computer, cable TV, and telephone, you won't be able to participate in the system at all.'"

'37 Chevy Gypsy Wagon

In 1972, Jack Fulton and I photographed this beautiful converted flatbed truck for Shelter. Joaquin de la Luz had built it and he and his wife Gypsy and their three kids (Heather, Bear, and Serena) travelled and lived in it for about five years. Last year we heard from Serena, now grown up, about her experiences living in the wagon:

MY EARLIEST MEMORIES of the Gypsy Wagon begin when I was three or four years old. At that point, our family had settled down in a little house on the Klamath River, in Northern California. We had all moved out of the Gypsy Wagon but I really missed it. I remember begging my mom and dad to let me use it as my bedroom. Luckily for me, my parents were such free spirits that they could really relate to my independence. The wagon became my room. I have memories of kissing my parents goodnight, leaving the house, and walking to my own little Gypsy Wagon. I had a huge doll that my mom had made for me, named "Howdy Doody." She made it out of vintage dress fabric, with old mother-of-pearl buttons for the eyes and mouth. Each night, I'd hoist Howdy Doody over my shoulder (he was bigger than me) and off we'd go. I loved the coziness I felt each night as I climbed into my bed. I remember the beautiful hand construction of the wagon, the texture of the wood, the hinges, and the little window above my bed. Everything about it was so warm. I think what made it so special was that it was filled with good intentions. My parents set out in the Gypsy Wagon because they were peaceful people. Their travels always had the purpose of happiness. The wagon was constructed almost entirely of other people's discarded junk. My father's creativity soared as he built it, and my mother made it a home. To this day, I really appreciate the warmth of simple things like old fabric and rusty metal. This is my history, as a child of free spirits with peace as their purpose. I wouldn't trade it for anything.

Rolling Homes

AN OUT-OF-PRINT classic on handmade mobile living in the U.S.A., *Rolling Homes: Handmade Houses on Wheels*, by photographer Jane Lindz, was published in 1979. Here are four photos from the book.

Pam built "Mañana" to ride on her '52 Dodge pickup. Design inspiration for the 6' × 6½' convertible cottage came from doll houses and canvas tents as well as gypsy and Conestoga wagons.

Gypsy trailer, Eugene, Oregon

The housetruckers build with materials they find at garage sales, flea markets, estate auctions, secondhand stores and demolition sites. Recycling materials decreases the cost of remodeling and increases the variety and character of their homes. Old and new objects, romantic and practical approaches combine to create a patchwork effect that recalls memories of the past and enriches fantasies of future adventures on the road.
—Rolling Homes

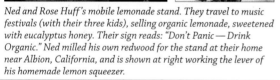

Rolling Organic Lemonade Stand

Ned and Rose Huff's mobile lemonade stand. They travel to music festivals (with their three kids), selling organic lemonade, sweetened with eucalyptus honey. Their sign reads: "Don't Panic — Drink Organic." Ned milled his own redwood for the stand at their home near Albion, California, and is shown at right working the lever of his homemade lemon squeezer.

Early SEER solar conference in Willits. Good vibes, good music, good food in the days before Real Goods tanked and SEER got corporatized.

Ned squeezing lemons with homemade press

Ananda's Gypsy Wagon

ANANDA BRADY built this gypsy wagon in the early '80s on the chassis of a '55 Chevy ½-ton pickup. It was built to be pulled by horses. He patterned it after gypsy wagons, but says it's "... a simulation, not a replica." He and his wife Cilla, and son Leon lived in it for two years. At the time these photos were taken, Flower Sierra was living in it.

1923 Model T Ford Camper/Bluegrass Show

Rod Cathcart

ROD CATHCART AND BOB BARKWILL tour the U.S. and Canada with this Model T camper (Rod found it in a Nebraska barn in the '70s), playing bluegrass at RV festivals. They tow it around on a flatbed trailer and seem to be having the time of their lives. "We live a dream," says Rod, and they call the truck Dream Camper. Sometimes they set up near a fancy RV park, with towed outhouse, put wash on the line, and the park will pay them $50 to move.

 www.dreamcamper.com

Truck house in the Pyrenees region, France (www.archilibre.com)

Truck parked in Santa Cruz, California, at a music festival. Owned by Peter Seydel Vincent, a wind-sculpture artist from Sacramento, California.

Jim Macey's Portable Cabin

IN 1980, Jim Macey built this portable 8' × 20' cabin. Under the floor joists there are two 8' lengths of 4" steel pipe that Jim uses to jack the building up. He then slides a specially built tandem-axle trailer underneath to move it. A raised skylight runs the length of the building and gives it a "caboose" look. Note the "eyebrow" flashing over the end window.

Log cabin on wheels photographed by Peter Wiley in Oakridge, Oregon in 2002

 www.mrsharkey.com
www.oldwoodies.com
www.rv-busconversions.com

Bread Van Home

HOWARD'S bread van, converted to a cozy, compact bachelor's home, in the desert near Death Valley, California. The aesthetics of simplicity.

This heavy-duty camping trailer was parked at a beach northwest of La Paz, Baja California, Mexico. The owner was not around that day.

Short school bus

Canvas tent built on wooden frame by Ole and Manya Wik, shown here on a barge in Glacier Bay National Monument, Alaska

Variety on the Road

Complex camper shell spotted in San Francisco one day in the '80s

Pic on the web that was too good to pass up. We don't know where it is, but it sure is a witty design.

Nomadic cappuccino cafe, with solar-powered lights, based in Gualala, California

'53 Road Van

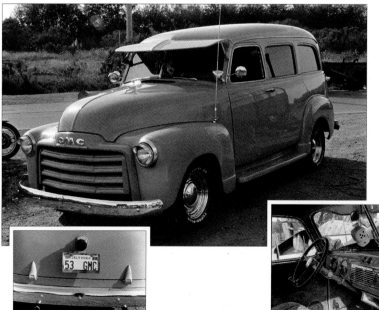

I RAN INTO Desi Whitman and his immaculate panel truck in Arcata, California in the '80s. He was a carpenter and had just taken a month's trip from Baja up to Washington on the coast, then to New Mexico for that year's Rainbow Festival. He carried his tools along and did remodeling to pick up road money. He slept in the back of the truck. His wife lived in L.A. and would fly to meet him on the road once in a while. Here he's hanging out with his friend Doug Connor of Arcata Scrap and Salvage.

more...

Air Camping

On a camping trip to the Sierras in the early '90s, I drove down a dirt road to Bowman Lake to take a swim. There, parked on a flat overlooking the lake was a great-looking camping rig. It consisted of a tough little Toyota jeep equipped with all sorts of rough country gear (such as the aluminum panels shown mounted on the side for getting out of sand or mud). On top of the jeep was a tent, with a ladder going up to what amounted to a second-story sleeping loft. The owner was down swimming in the lake and the tent looked cool and inviting, with mosquito netting hanging in the entrance. I was fascinated. This way you were up off the ground, no worry about snakes or scorpions in the desert, with cool breezes and a great view. After a lot of research I located the manufacturer, Air Camping, in Italy, and bought one.

I've used mine for over ten years now, most extensively in Baja California (see p. 169). It folds up into a compact roof-top unit. When I stop at night I take off the cover and I pull open the folded-up section, which automatically opens the tent. There's a ladder that holds up the cantilevered section and a built-in mattress. I have a regular bed with sheets, blankets (if necessary), and my own pillow. Comfortable!

These units are not cheap; expect to pay at least $800–$1000 with shipping, but for serious campers, they can be worth the investment. (They can also be put on top of passenger cars.) Here are websites for more info:

www.loftyshelters.com
www.bimo.com/skydome1.htm
www.autocamp.de

Surfer Tod's truck held together by decals

Paolo's Pinzgauer all-terrain vehicle. www.pinzgauer.com

Shortboard/longboard beach vehicle

The unobtrusive *Rental Car Camper (Tips from John Welles): Most foreign cars (e.g., Honda, Toyota, Datsun) have seats that fold down to 30° (American cars do not) — perfect for sleeping. It's important to be inconspicuous. You can often sleep in residential areas with no window curtains — you're out of sight. Another essential: a trunk that cannot be opened from inside the car with a latch (after say, smashing a window), for leaving valuables while you're out exploring.*

Shade from the Sun

Hey desert rats! What you need out there is shade. It makes life comfortable in hot weather, and bearable when it's superhot. This is an ingenious, lightweight, and cheap shelter utilizing a 1″ electrical conduit frame connected with special fittings and wingnuts, with a silver awning attached with "ball bungees." I had a 12′ × 14′ gable roof tarp for years in Baja. It cost about $200, and folded up in a Yakima Rocket Box on the truck's camper shell. It took me about an hour to set it up. You can also get side panels that cut strong winds down to a tolerable breeze. Three places to buy them (I bought mine from Jenkins and they were great to deal with):

Jenkins Crafted Canopies, Costa Mesa, CA, www.jccshade.com

Thomas Tarps, 375 Helroy Way, Arroyo Grande, CA, 93420, 805-489-1737

Tarps Plus: http://www.tarpsplus.com

Capitalist Pig ▲

Capitalist pig SUV (Slob Utility Vehicle), gets about 8 mpg — flagrant waste of precious resources. Will probably never get off-road. Parked in Mill Valley, California, in 2003. Or check out the Cadillac Escalade for another gross vehicle. Embarrassingly uncool!

Liferaft for 4–25 people: Switlik Parachute Co., Trenton, NJ www.switlik.com   609-587-3300

Housetruck parked in Santa Cruz, California

Homemade Curved Camper

Ed O'Connor, a sheetmetal worker, built this curved camper shell in the '70s. Inside, the curved roof gives more headroom and a more spacious feeling than a flat roof. His son Brendan, shown here, now has the camper on his truck. (Surfers sometimes use a 4" pipe mounted like this and filled with water, for post-surf showers; or, it can be used to carry fishing rods.)

Armored Dodge Powerwagon

Mr. & Mrs. H. L. Baggett's 1948 "Armored Field Headquarters," built on a 1948 Dodge Powerwagon chassis. It weighed 41,000 pounds, had a 3"-thick mineshield under the chassis, a 700 cu. in. flathead six engine, had 4-wheel drive, three 50-gallon gas tanks (it got ¾ miles per gallon), solid rubber tires, an 8-ton winch with its own 3-speed gearbox, and all instructions in it were in Spanish. It had a gas/wood cookstove. In 1993, Mr. Baggett wrote us:

Hi Folks! It has been 20 years since we bought your magazine, Shelter, *and we would like to know how the past 20 years has treated ya! We have been through many cars, buses, a couple of boats (one built in 1680!), a couple of tents, and a bridge or two! Followed the grain harvests, been in the oil fields of Louisiana, and worked with the Carnies and Circuses . . . . The missus and I have been married 26 years, and still going strong.*
                                                                    —H. L. Baggett

L.A. Filmmakers

www.lafilmmakers.org
contact@lafco.tv
310-574-4733

Alfonso Gordillo and Tao Ruspoli, Europeans in their mid-20s, bought a 1985 Chevrolet Bluebird school bus on eBay for $3000, had it outfitted in Los Angeles, and set off on a movie-making tour across America in September, 2001.

This operation, and the bus, are called L.A. Filmmakers. Their mission is to ". . . travel and make our own films and help people make films who don't have the equipment to do so . . ."

The bus is set up with state-of-the-art Macintosh equipment, flat monitors, digital video cameras, a projector, editing equipment, and a major sound system. Couches fold down into beds for sleeping. Alfonso and Tao are multilingual (6 and 3½ languages, respectively) and are paying for the trip by working as translators on their cell phones. They make about $2500 a month each by doing 8-hour shifts whenever they get to an area that has cell-phone connectivity. They typically will be interpreting for two people, one speaking Spanish, the other English, on the line in a three-way conversation.

The bus has a wonderful library of both books and films on VCR and DVD. When I ran across them (in Pt. Reyes Station in Marin County, Northern California), they along with their travelling companion Roger Webster, had just come from the Burning Man Festival in the Nevada desert, and were heading up the California coast to British Columbia. Between them they had five cell phones. These days the bus is parked in Venice, California, where there are workshops and meetings for filmmakers, photographers, and other artists.

Alfonso and Tao

Handmade Housetrucks and Housebuses
Roger D. Beck

ROGER BECK built his first housetruck in 1969 and spent several years on the road as a traveling artist, making wire jewelry that he sold at crafts fairs. Many of the vehicles shown here were owned by Roger's friends; they often travelled in groups, helped each other building, did crafts fairs together, and ". . . there were also times when we just kicked back and enjoyed a simple life." Roger began shooting photos of houseboats and housebuses. "I originally took these pictures for my own enjoyment and put them in a photo album." Then in 2002, with encouragement from a friend, he published *Some Turtles Have Nice Shells*, a 200-page handmade color compendium of soulful road living in America. On these two pages are photos from the book.

Below is Roger's website (to get this book), as well as Sharkey's huge website of house vehicles:

www.housetrucks.com
www.mrsharkey.com

In the '70s, Michael was on the road in this housetruck (above three photos). When he hit a new town, he'd find the fanciest supermarket, drive past the front window a few times (so everyone could see it), then park in the back of the parking lot (so as not to be in the way), and open up to the public. People would pay 50¢ admission, walk through the truck, and buy postcards (of the truck, of course). At the end of the day he'd have enough cash to buy groceries and gas, and still have some money left over. Great gig!

Roger Beck's 1951 Federal five-ton is his fourth housetruck. He carried a 1940s Whizzer motorbike along as a "dingy," and sold jewelry at art fairs. Roger says "I now dream about building number five and being back on the road again and living a simpler lifestyle!"

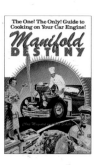

Grant and Elissa looking at a copy of Shelter

Hi-tech rock climbing gear

Bedroom in 30 seconds

Outdoor Adventures

Grant Cahill and Elissa Vaessen spend a good part of the year camping, mountain biking, kayaking, and rock climbing. Grant, it turns out, is the traffic manager for a large company in Vancouver, B.C. (Canada) and says he has a deal with the owner; he works hard seven months a year: long hours, handling the shipping, driving a forklift, etc. "For seven months, he owns my soul — the other five months, I play." And he adds, "Mind you, he pays me exceptionally well."

Their rig is a '98 Ford Ranger with a shell that cost $100. Last year they went through 14 U.S. states and two Canadian provinces, took a tandem fiberglass kayak out to the Gulf Islands (between Vancouver Island and the mainland), rode their bikes along the Scorpion Trail in B.C., and went bouldering and rock climbing.

Flip-Pac

Steve and Sondra Winslow and their 4×4 Toyota Tacoma pickup truck fitted out with a Flip-Pac camper shell. Steve says it has a torsion spring and pops up in 30 seconds. Costs around $3000. Steve and Sondra run day-long, low-cost Colorado River rafting trips.

 www.flippac.com
www.coloradowhitewaterrafting.com

Patagonia to Alaska on Muscle Power

Last year I drove past two cyclists with heavily laden bicycles stopped by the side of Highway One. Attached to the back of one of the bikes was a high-tech-looking little trailer piled with stuff, and a sign on it saying "Patagonia to Alaska — No War!" Whoa! This looked interesting; I turned around and came back to meet Silvia Monja and Alejandro (Baldy) Barreiro, two charming, tough, resourceful and amazingly fit Argentinians who had been on the road over a year.

They left Patagonia a year ago and had so far pedaled over 12,000 miles. They were carrying their shelter on the bikes; Saturday night it had rained hard and they had a tent and sleeping bags to dry out, so I invited them to come to our house. They hung up their gear and I made them omelets and toast. They ate and ate! They'd been married seven years, and the bike ride was Sylvia's idea. It's amazing to meet people with this kind of spirit. They'd just been through South and Central America, then Mexico and now into California — all on human power. People in all countries would invite them in, give them food and a place to sleep. They worked four years to save money for the trip. Things were fine for about a year. Then in December, the Argentine economy collapsed and they were stranded in Mexico without funds. They started going to Mexican firehouses asking if they could pitch their tent nearby and they usually ended up with a place to stay and food. When they entered the U.S. at Nogales they were down to $4, and they found that guys at American firehouses likewise provided food and shelter.

The next morning they asked me to come with them on my bike for a few days and I was tempted, but alas too busy, but I rode to the edge of town with them. As we headed around the lagoon, I decided to speed up and get far enough ahead so I could take a photo. I pedaled real hard for several hundred yards and looked back and there was Baldy right alongside me, smiling, pulling at least 125 pounds of gear.

We shot the pictures, and as they pulled onto Highway One, heading for Seattle, then Alaska, Baldy reached back and flipped on his solar radio to a classical music station.

Their website (in Spanish): with spectacular photos of them going though deserts, rain forests, valleys, the Andes, at Machu Picchu:

www.bicitrips.com.ar

Books for the Road

Manifold Destiny, by Chris Maynard and Bill Scheller, 1998, Villard, New York
Poached fish Pontiac, pickup ham steak, stuffed cabbage — cook it on the road and on your engine! Why not? This classic book gives you 35 tested recipes, tells you how to wrap it in foil, where to place it on different car engines, and how long to cook it. Out-of-print, but I got a copy through Amazon.

Car Living Your Way, by A. Jane Heim, 1995. Touchstone Adventures, P.O. Box 177, Paw Paw, IL 61653
A unique book about the culture of short-term and long-term car living. Explains how to live comfortably in a car if you want to (or have to). There are over 100 stories from all sorts of people living in cars, and tips and tricks, as well as a lot of reference info. A lot of the advice is aimed at single women living in cars. "For some of you, it will take years to get things ready to go. Start simple, start slow, but go. 'Til then, as the Irish say, 'May the wind be at your back and the sun shine gently on you.'"

Roadside America, by Jack Barth, Doug Kirby, Ken Smith, and Mike Wilkins, 1986, Fireside/Simon & Schuster
A wacky, chatty, fact-filled book on America's roadside wonders and amusements.

A large info-filled website on house-buses and housetrucks:

 www.mrsharkey.com

LIVING LIGHTLY

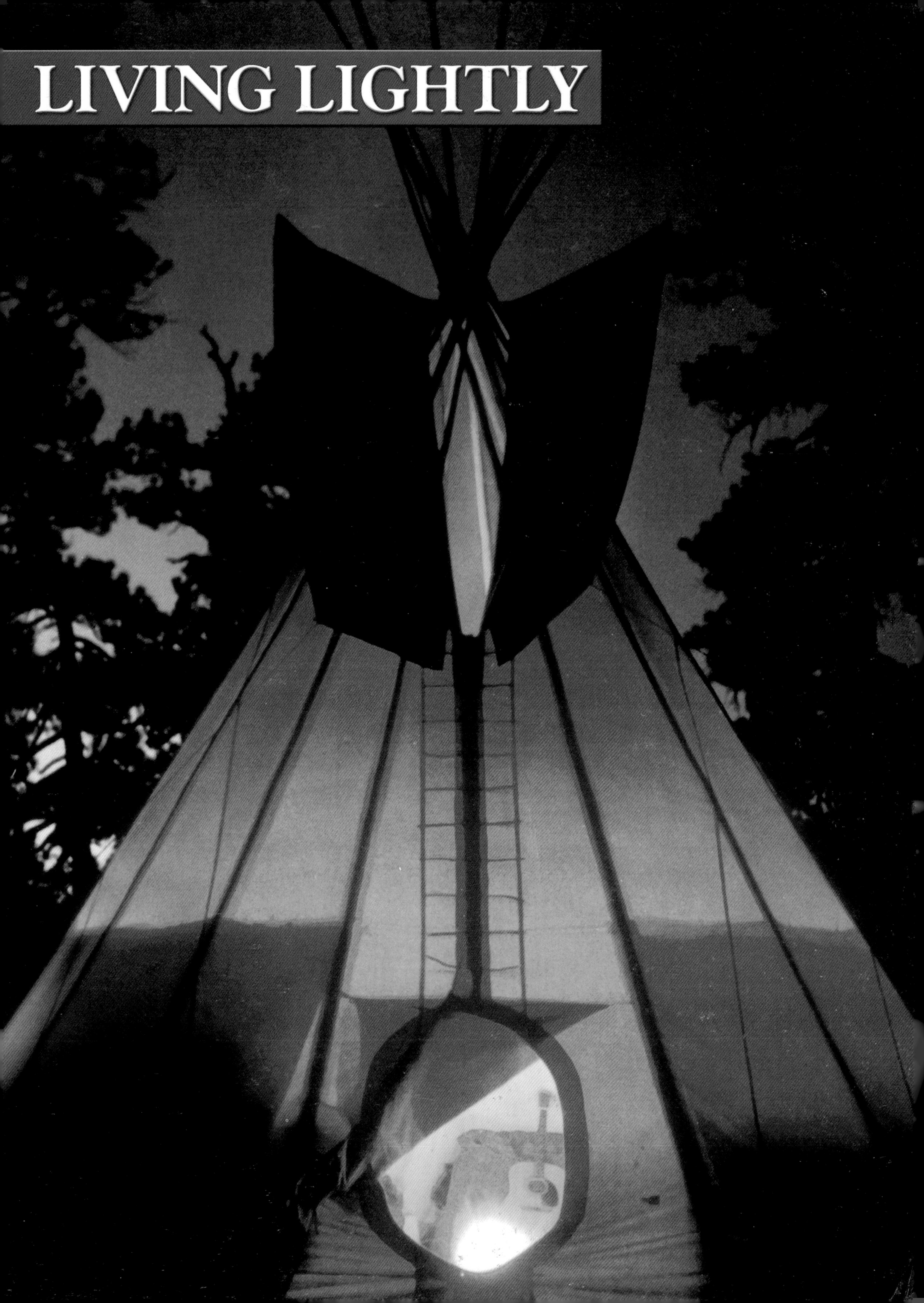

PERPETUAL CAMPING

In the early '90s I received a little mimeo newsletter called Message Post *in the mail. It was tightly packed with tiny print and had tons of info on what Holly and Bert Davis called* Living on Land Lightly. *Holly had written a great reply to one of my objections to domes, i.e., that homes are not portable. Turns out that Holly and Bert and a bunch of uncounted nomads out there do utilize movable homes, and* Message Post *(now called* Dwelling Portably*) is their low-tech, fact-filled newsgroup-by-U.S.-mail. (No website here!) At right is a description by Holly of how they got started doing this, and excerpts from their publications about lightweight shelter in the woods.*

Dwelling Portably is $1 an issue (can you believe it?), add 50¢ if sending check or M.O. less than $6. Hell, send these guys $10, they are doing something unique here — talk about low consumption of planetary resources! Dwelling Portably, POB 190-hwk, Philomath, OR 97370

WHILE QUITE YOUNG, Bert and I decided (separately, before we even met) that buying property was foolish. You can't really own land; the government owns it and can kick you off any time you do something that any of dozens of government agencies disapprove of. Or, if you rent, you pay for the hassles and anguish of those who do buy.

We also noticed that much land, especially in the West, was not used (by humans) or was used infrequently. That inspired us to become perpetual campers: living in a place while it was desirable; moving on when conditions changed.

Unfortunately, most manufactured equipment is intended for recreationalists who camp only a few days a year, mostly during summer. Even "four-season" tents: though you may *survive* in them year around, you probably won't be very comfortable. So presently, to dwell portably for long periods *in comfort*, requires much do-it-yourself.

In *Dwelling Portably*, doers report on what works and doesn't, ask questions, and offer advice. Though readers vary widely in how and where they dwell, most live as simply as they can comfortably. Issues vary: some have much about vehicular dwellings and little about backpackables or wickiups. Or vice versa. So, for a broad sampling, order several back issues. Our prices encourage that (and pass on postage savings): 1/$1; 6/$5; 13/$10; 30/$20.

Brief history. In the late 1970s, a network of portable dwellers developed in the Northwest. *Dwelling Portably* (original name *Message Post*) began in 1980, produced mostly by Hank and Barb Schultz. About that time, Bert and I learned of the network and moved to Oregon. We began caretaking *Dwelling Portably* during summers while Hank and Barb were away working. Since 1987 we have been year-around managers. There have been two or three issues per year.

Bert and I have built portable dwellings that are as comfortable as houses. In some ways they are more convenient, e.g., because they are small and well insulated, our body heat keeps them warm during winters — avoiding the labor, mess, pollution, and hazard of a heating stove.

–Holly & Bert Davis

The Hillodge: Hillside Living

SIMPLE DWELLING IS COMFORTABLE IN MOST WEATHER

The Hillodge is designed for portable living year around on steep slopes. A clear front wall, combined

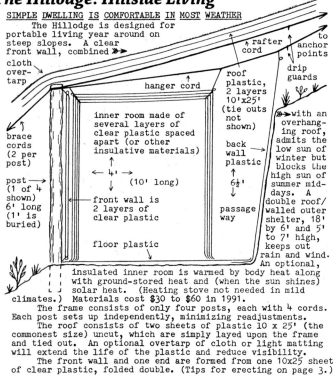

with an overhanging roof, admits the low sun of winter but blocks the high sun of summer mid-days. A double roof/walled outer shelter, 18' by 6' and 5' to 7' high, keeps out rain and wind. An optional, insulated inner room is warmed by body heat along with ground-stored heat and (when the sun shines) solar heat. (Heating stove not needed in mild climates.) Materials cost $30 to $60 in 1991.

The frame consists of only four posts, each with 4 cords. Each post sets up independently, minimizing readjustments.

The roof consists of two sheets of plastic 10 x 25' (the commonest size) uncut, which are simply layed upon the frame and tied out. An optional overtarp of cloth or light matting will extend the life of the plastic and reduce visibility.

The front wall and one end are formed from one 10x25 sheet of clear plastic, folded double. (Tips for erecting on page 3.)

Storage and Sleeping

I camp on an island in Maine.

I live there 7 months a year in a large green tent. The tent sets on a poly tarp that covers a thick mat of twigs and leaves. That keeps the tent's inside drier, I hope. (The site is wet.)

I use 4 big plastic containers to store food, tools, clothes, and

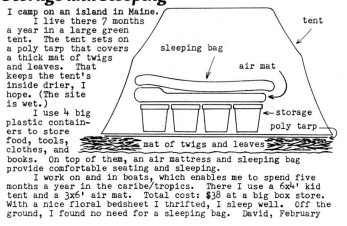

books. On top of them, an air mattress and sleeping bag provide comfortable seating and sleeping.

I work on and in boats, which enables me to spend five months a year in the caribe/tropics. There I use a 6x4' kid tent and a 3x6' air mat. Total cost: $38 at a big box store. With a nice floral bedsheet I thrifted, I sleep well. Off the ground, I found no need for a sleeping bag. David, February

TIPS FOR ERECTING A HILLODGE

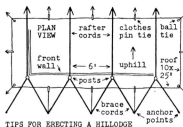

I prefer southerly slopes during fall and winter, and easterly slopes during spring and summer (for morning warmth).

For a site, choose a 10 by 25 foot area clear of high bushes or trees. Uphill must either be steep, or else have fair size trees to provide elevated anchor points for the rafter cords. (Avoid large trees dangerously close.) Downhill ideally has no trees or tall bushes close enough to block light. Level and smooth a 6' by 18' terrace.

Only shallow post holes are needed, because cords will brace the posts. I round and smooth the top of each post so that it won't puncture the roof plastic. Carving a groove close to the top helps to keep the rafter cords in place.

Seldom are anchor points for rafter cords ideally located. I use whatever are available and then reposition each rafter cord with an auxilliary cord pulling sideways. I run separate cords for hanging things on, rather than use the rafter cords and sag them.

If the rafter cords must fasten to trees (because uphill isn't steep), I tie to the bases of branches (rather than around trunks, to avoid bruising), and then brace to logs, roots, or bases of bushes (to avoid bending the trees).

For the roof, black plastic will out-last clear, especially if no cloth over-tarp. To tie out the plastic, I use 1½-2" diameter ball ties on the corners, but spring-clamp clothespins on the sides (where ball ties might form rain-collecting pockets). Where clothes-pinning, I roll the edge of the plastic a few times around a straight twig, 1" diameter a few inches long; clamp to the roll; then tighten the grip by wrapping a narrow rubber strap around the clothes pin. I tie the tie-out cord through the clothespin's spring.

I stretch the roof's upper layer tauter than the lower layer, to form ±1" air spaces between. For an overtarp, I use any old drab-colored cloth, pieced together. Because the plastic is water-proof, the cloth need not be.

On the uphill side, just inward from the edge of the plastic, I wrap a piece of string or strap around each cord, so that any water running along the cord will drip off there rather than inside.

April 1992 Page 3

I ball-tie the front wall plastic to the posts, bottom as well as top (rather than just set rocks/dirt on the bottom), for reliability and to save time moving.

I cover the back wall and floor with plastic, to block moisture and confine any loose dirt or leaves. Odd/holed/salvaged pieces may be used, overlapping as necessary. For comfort and appearance I add mats, rugs or drapes.

An insulated inner room can be built in various ways out of many different materials: anything that can be kept in position and will provide several inches of loft. We have used our Hillodge only during spring so far, and thus have not needed an inner room. When I build one I plan to use several layers of clear plastic for sides and top, suspending them ½" to 1" apart with a succession of ball-ties-with-tails. Such can be made by tying a bulky knot in a short length of cord. To erect, wrap the top layer of plastic over the bulky knot and tie the suspension cord around the base of the bundle, letting the tail hang through. Then tie the tail to the next layer of plastic in a similar manner. Etc. How I build the ends of the inner room will depend on the materials at hand.

If a 6 by 18' Hillodge is not roomy enough, rather than enlarging, I would build additional shelters near by. (Leveling a site for a Hillodge twice as big would disturb 8 times as much dirt.)

Limitations of the Hillodge: The roof is not steep enough to shed snow, nor strong enough to support an accumulation, and therefore should not be left up and untended when heavy snow is likely. The plastic roof and front wall may not take gales. The frame is not well suited for areas which are both flat and treeless. The outer shell includes seams that will admit insects unless carefully chinked. (A bug net can be hung in inner room.)

Tho the frame, roof, and most walls are simple, the end closures are irregu-lar - and are left for the builder to work out. (Detailed instructions would not only fill many pages, but might take longer to read and understand than to invent for yourself.)

Comparing the Hillodge with Wanda's Tleanto (in Sept'85 and Sept'86 MP):

The Tleanto's roof is steeper and stronger, and thus can shed snow or support an accumulation. The frame is self supporting, not requiring high anchor points; thus is not restricted to steep hillsides or edges of groves. The front wall is higher, providing more solar heating and more light inside, especially high up. If south facing, a Tleanto does not need a cloth overtarp to prevent reflections of the sun.

The Hillodge's frame (only 4 posts) is much simpler and lighter. The rafters are cords instead of poles, and thus are not hazardous if they break. The roof slopes the same direction as the land, and thus is less intrusive and less buffeted by wind. The inner room can be placed at the front, with a full-height passageway behind it. For the same width of terrace, it has more floor space. B&H

The Snugiup: $20 Shelter

A naturally warm-and-cool shelter that needs no poles.

The Snugiup is a small enclosed dugout formed entirely of earth and plastic. Thus it can be built even in areas that lack timber such as deserts and brushlands.

The roof and front wall are insulated by air spaces between the layers. The other three walls and the floor are mostly soil, and act as natural, automatic heaters-coolers; absorbing heat on warm days and giving it back during cold spells. In western Oregon winters: when occupied, inside is typically 65°; seldom below 50° even when 15°F outside. (We've not had any weather colder than that to test it.)

The roof is flush with the ground, which minimizes wind forces and facilitates concealment.

The roof will not support much snow, but may allow snow to slide off, depending on steepness of roof and the type of snow.

Materials cost $10 to $20. Erection requires an hour or so, NOT counting site preparation which may take several days (but can be done in advance of use).

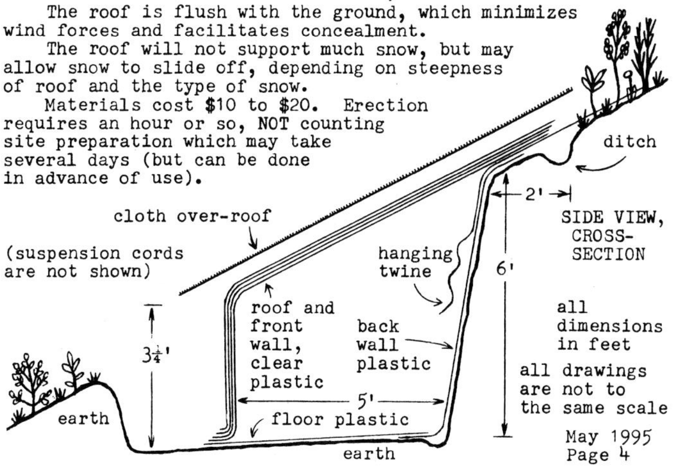
ditch
cloth over-roof
2'
SIDE VIEW,
CROSS-
SECTION
(suspension cords
are not shown)
hanging
twine
6'
roof and
front
wall,
clear
plastic
back
wall
plastic
all
dimensions
in feet
3¼'
all drawings
are not to
the same scale
5'
earth
floor plastic
earth
May 1995
Page 4

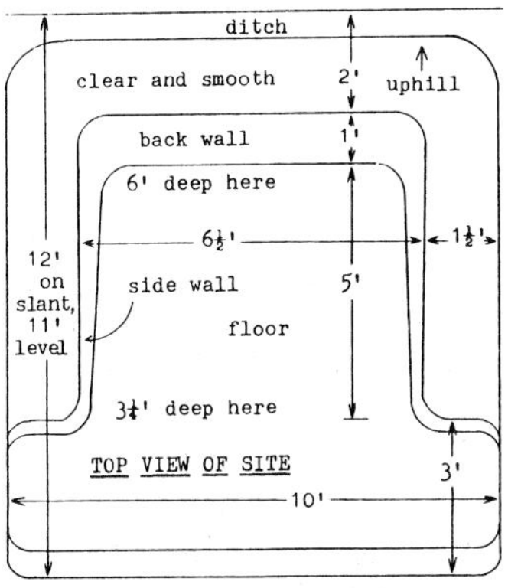
ditch
clear and smooth
2'
uphill
back wall
1'
6' deep here
6½'
1½'
12' on slant, 11' level
side wall
5'
floor
3¼' deep here
TOP VIEW OF SITE
3'
10'

Suggestions for building a Snugiup.

Construction is easiest on a steep slope. For winter, a south-facing slope will be brightest and warmest. For summer, an east-facing slope will warm during mornings and cool during afternoons. An ideal site has much foliage overhead, but little foliage downhill.

The Snugiup's walls require firm soil that will not collapse. Avoid soils with much sand or gravel. Roots are desirable for reinforcement though they will slow digging. Digging is usually easiest when the soil is somewhat moist. Early spring may be a good time to prepare a site for use the following winter; and allows time for soil instabilit to maybe show up. CAUTION: build and live in a Snugiup (or anything) at your own risk! We make no guarantees and assume no responsibilities or liabilities.

For a site also consider: can the dirt be disposed of easily and inconspicuously

The illustrations assume a uniform slope that rises one foot for every two feet level distance. Most slopes don't. Furthermore, less digging is needed where a depression or irregularity exists. Therefore, modify to suit your site.

To prepare site, clear an area 10 by 12 feet. If much rain expected or if the soil does not drain well, shallowly ditch along the back (uphill) side of clearing.

Dig out the portion shown. Slant the walls, especially the back wall, to reduce risk of slumping. (Don't make the walls straight up and down.) Round all corners. Remove any rocks protruding from walls and tightly pack their holes with moist dirt. Cut off roots flush with wall. (Don't try to pull them out.)

Slope the floor slightly downhill. If ground water expected, ditch around.

If the site is left vacant for long, the walls will be eroded less by frost or dryness if covered with plastic.

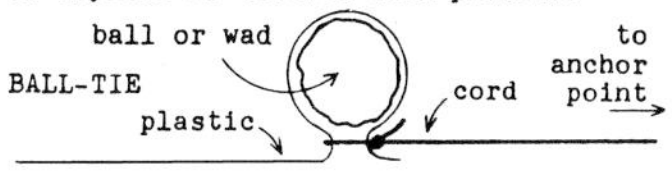
ball or wad
BALL-TIE
plastic
cord
to
anchor
point

To erect, hang pieces of white or clear plastic over the walls, holding in place with ball-ties going to bushes/logs/rocks beyond the cleared area.

Hang lengths of twine on the back wall (for suspending light-weight items when the Snugiup is occupied).

Form the roof and front wall from a piece of clear plastic 10 by 12½ feet. (If the plastic comes in rolls 10 by 25 feet, simply cut in half.) Anchor with ball-ties along sides and back. (Do not put ball-ties along front of roof, because they would form puddles.) Let excess length lay on the floor.

Suspend a second 10x12½ piece of clear plastic ½" to 1" outward of the first. If possible, tie its anchor cords to different bushes than hold the first layer, so that tension on one layer does not slacken the other layer.

Add additional layers as desired for insulation. (For summer use only, two layers are probably enough; for winter, four or more layers may be desirable.)

Cover floor with plastic. Add rugs, drapes as desired for comfort or decor.

Entry is by raising the front wall plastic and ducking under.

Where not much foliage is overhead, if shade during summer or concealment from air are desired, form an over-roof from appropriately-colored cloth 10½x10½ (need not be waterproof). Suspend it about 6" above the roof plastic. Put its front (downhill) edge so as to block direct sunshine during summer when the sun is high, but to admit sun during winter when the sun is low in the sky.

If extremely hot and sunny, add a second over-roof of white or reflective material, suspending it 6" above the plastic and 6" below first over-roof.

The Snugiup is small inside: floor 5 by 6 ft; height 3 to 6 ft. Therefore most belongings must be kept outside in stashes or under tarps when not in use.

Wintering in a Snugiup.

In mid November I built a shelter like the one in May'95 DP (also shown in 1995-96 Summary-Index) and have been living in it since. I used six plastic liners and two roofs. The top roof is cloth; the under roof is clear plastic. Most days the shelter stays warm enough to wear just a T-shirt and watch cap, or nothing. During one freeze I had to wear more, but stayed comfortable. This is with no stove. The shelter has stayed dry except for a little condensation on the front wall.

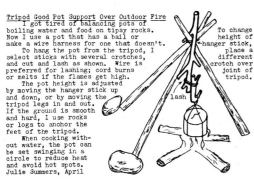

Tripod Good Pot Support Over Outdoor Fire

I got tired of balancing pots of boiling water and food on tipsy rocks. Now I use a pot that has a bail or make a wire harness for one that doesn't.

To hang the pot from the tripod, I select sticks with several crotches, and cut and lash as shown. Wire is preferred for lashing; cord burns or melts if the flames get high.

The pot height is adjusted by moving the hanger stick up and down, or by moving the tripod legs in and out. If the ground is smooth and hard, I use rocks or logs to anchor the feet of the tripod.

When cooking without water, the pot can be set swinging in a circle to reduce heat and avoid hot spots. Julie Summers, April

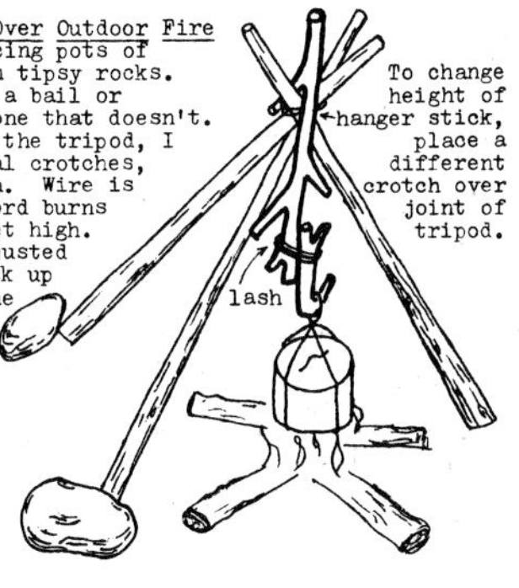
To change
height of
hanger stick,
place a
different
crotch over
joint of
tripod.
lash

Winter in a Tleanto

Preparing to Winter in a Tleanto

We plan to camp in the mountains this winter. We have most supplies cached and are now preparing the site and cutting the poles. We'll wait until late fall to set up and move in.

The Tleanto is a lean-to within a plastic tent. Like the twipi it has two walls (actually 2½ counting the fly).

Our site has a SSW slope (would prefer S or SE but didn't find one with the other nice attributes).

Craig dug two terraces. The larger lower one will be the floor, the upper one the rear support for the roof poles.

We will cover the ceiling with plastic in back but with old sheets in front to let moisture out. On top we will pile leaves for insulation (moss might be better but not plentiful).

The front wall will be four layers of clear plastic spaced apart with some strips of bubbly plastic we found (used for padding when shipping). The end walls will be similar except that the plastic of the east wall will have one corner anchored with elastic straps so that we can squeeze between it and the corner post to go in and out.

The tent will be shaped to match the lean-to but otherwise similar to the Woodland Tent (in LLL catalog), with plastic gathered together to close one end, and entrance covers (cloth top, plastic bottom) over the other.

The lean-to will be 12' long, 6' wide, 5' high in front, 2' high in back. The tent will be 24' long and extends beyond the lean-to to provide an outer room that will be sheltered from the wet but not insulated. Materials, mostly plastic, cost us less than $50. For cooking we have a gas stove (RV salvage, also tank and regulator). The kitchen will be in the outer room, at first anyway, but may try moving it inside the lean-to portion on cold days.

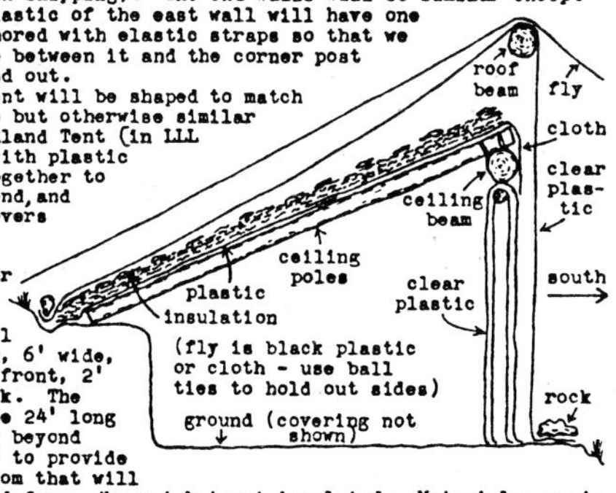
roof
beam
fly
cloth
clear
plas-
tic
ceiling
beam
ceiling
poles
plastic
insulation
clear
plastic
south
(fly is black plastic
or cloth - use ball
ties to hold out sides)
ground (covering not
shown)
rock

We have spent three summers and parts of springs camped around here. This will be our first winter. Suggestions are welcome. Our last mail run will be early November.
Wanda, California, July

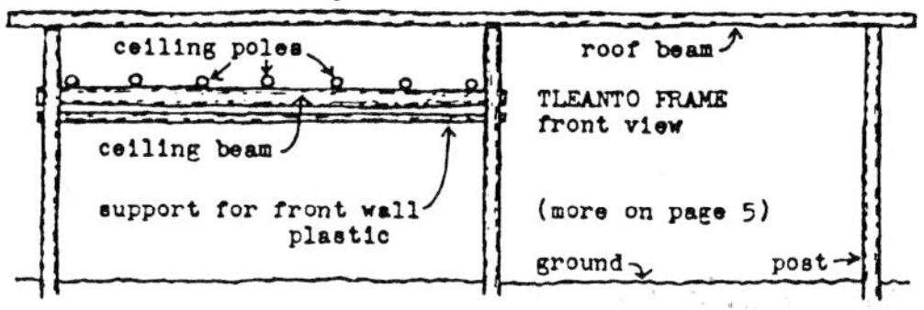
ceiling poles
roof beam
TLEANTO FRAME
front view
ceiling beam
support for front wall
plastic
(more on page 5)
ground
post

My Family and I Now Live in TWO Tleantos

The new one is presently about 400 miles south of the first one. We migrate between, not so much for winter warmth (the southern spot isn't much warmer then), as for a sunnier spring and to be closer to the couple whose kids we board.

The new tleanto is much like the original (Sept'85 & Sept '86 MP) except longer: 36' total. (Actually there are two separate frames joined end-to-end with covers overlapping.) It contains two insulated rooms plus some uninsulated areas.

Usually two adults and three small children live in it; on occasion two additional adults. During the coldest weather for warmth we all crowd together into one room. (One morning I measured 22°F outside, 30° within the outer shell, and 48° in our room.) We don't want a heating stove with the fire hazard, fumes, smoke, work. During milder weather we spread out. The tleantos are warmest where there is little wind.

For ceiling insulation, I replaced the leaves with sheets of flexible foam, which are quicker to put on and easier to keep in place. I am now gradually replacing the foam with bubble plastic (as we find it) which has not been bothered much by animals (whereas the foam becomes nests if not protected). We take down each tleanto when we leave for the season, to avoid mouse/rat infestation.

Another change: I angle the south wall with the top farther out than the bottom, so the sun doesn't reflect far. (Once we were careless and attracted a hunter who was on the slope below. Luckily, he was friendly.)

I have now lived in tleantos for most of five years and am quite satisfied with them. Wanda, California, August & Nov.

Jug Showers

I bathe each day with a #10 can half full of warm water.

I get these cans, free, from behind restaurants. I bathe in the afternoon sun or under the stars at night. I use a cup to dip from the can. I pour one cupful over my head. Then I spray a two-second squirt of diluted baby shampoo (ten parts water to one part shampoo) on my head. I use the rest of the water to rinse off, continuing to pour from the cup onto my head, letting the shampoo run down my body as it gets rinsed off. I've bathed like this since June 1998. Dana, TX 760, Apr

(Comments:) Sounds much like how we bathed before reading Julie Summers' paper, "The Simple Shower".

Advantages of a narrow-mouth plastic jug or bottle, compared to a cup: no repeated stooping to refill the cup; the jug's narrow mouth helps regulate the flow.

If I pour water only on my head, it does not wash/rinse my entire body. The water tends to form streams, instead of spreading over my skin, and it completely misses arm pits and crotch. So I soap and rub each part, working from my legs up; then rinse from the top down.

Bert and I have bathed with jug showers almost exclusively for 20 years. Even when sitting a house that has a piped shower, we use jugs so we can get the water temperature we want before wetting. Also, if there is enough privacy outside, we shower there, so we don't have to wash the shower stall.

During cold weather we now do as Wanda suggested in May 93 DP: soap and rub dry part-by-part while inside, so that we need be outside only long enough to rinse. Holly, OR 973, July

I REMOVE the cap from the jug when showering.

This illo, from June 87 DP and 87 Summary-Index, shows only one drop. Some readers assumed the cap is only loosened. That won't provide enough flow.

Also, I MOVE the jug a lot, tipping it up momentarily to get a brief gush while rubbing with the other hand. If held still, the jug drains too quickly. Getting the flow you want may take a little practice. Holly, Oregon 973, July

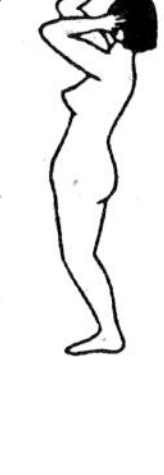

Plastic containers contaminate water or food.

Plasticizers leach out, which are toxic in the long run. Drinking water sold in plastic jugs may be LESS healthful than piped city water.
L.Smith, March

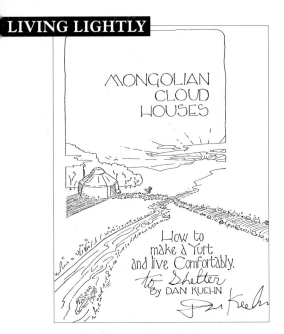

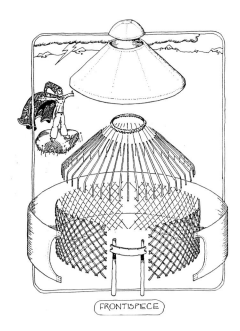

FRONTISPIECE

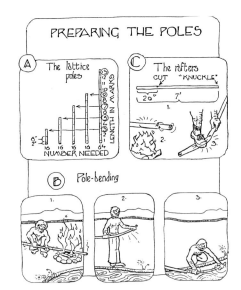

PREPARING THE POLES

MONGOLIAN CLOUD HOUSES

Dan Kuehn

In 1981, Dan Kuehn sent us an inscribed copy of his new book, *Mongolian Cloud Houses.* Although produced in the '80s, it was a book born of the '60s, with the energy and vitality of people who revered the earth and its resources and were dedicated to leaving as small a footprint and using as few resources as possible. At the time Dan was living in a 13'-diameter, 10'-tall homemade yurt in the woods. The book is a guide to building that yurt. The instructions are clear and the drawings are beautiful—informative and friendly.

Note: In 2006, Shelter Publications republished Dan's book, extensively updating and revising it—adding new techniques and materials, a resources section with suppliers and manufacturers, and a photo section of both Dan's original yurts and Mongolian *gers.*

www.shelterpub.com/_mongolian/MCH-book.html
www.mongoliancloudhouses.com

BONES

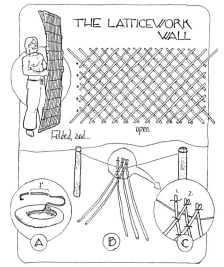

- *About 150 pieces bamboo (I prefer Johnson Grass), 10–15' long*
- *4 car tire innertubes (to cut into rubber bands)*
- *about 35 shoots young willow*
- *piece wood ¼" × 5½" × 41"*
- *two 7-foot poles*

THE LATTICEWORK WALL

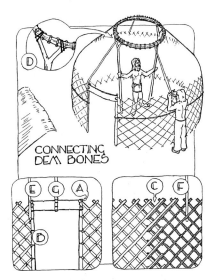

The latticework is made in 4 sections, each of which accordions up like a baby gate . . . you'll need about 700 rubberbands, which can be cut from car innertubes.

The only materials I had to pay for were canvas, needle and thread, safety pins, and waterproofing, for a total of less than $175. The rest was gleaned from the woods, backyards, and the local dump.

The first yurt I ever saw was most wondrous — I was so taken by the quality of the space that I decided to abandon my previous plan to make a tipi my home.

THE SMOKEHOLE RING

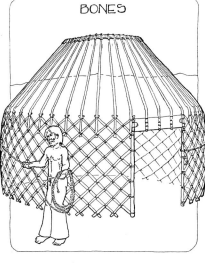

. . . cut 20 young green poles of willow or some other similar plant, about 12' long . . . a well-made smokehole ring is a work of art.

CONNECTING DEM BONES

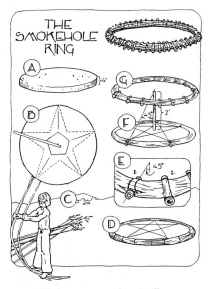

. . . It'd be handy for you to invite a friend to the yurt erection, particularly at the point of adding the smokehole ring and the first rafters. Other than that, you can do everything yourself.

WALL SKIN

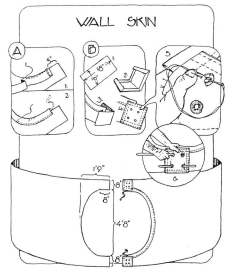

You can sew your own skin, either by hand or by machine. For this 13-foot-diameter yurt, you'll need 33 yards of 6-foot-wide canvas. I prefer 12 ounce untreated — it's both strong and natural.

ROOF SKIN

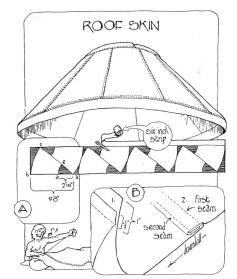

The main part of the Roof Skin is made by sewing 8 "pie pieces" into a cone shape.

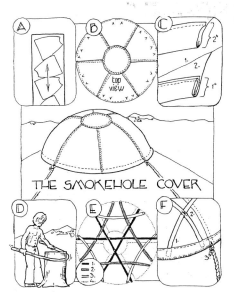

THE SMOKEHOLE COVER

The Smokehole Cover is a dome sewn like half a beachball, that is, 6 rounded triangular sections with a circle at the top.

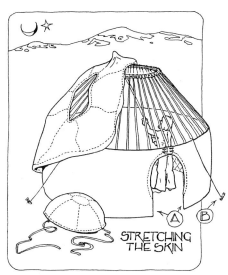

STRETCHING THE SKIN

Throw an edge of the skin up onto the roof frame so that some of it hangs in the smokehole ring. Then, with a 6'-long pole, you can maneuver the roof into place.

SKIN

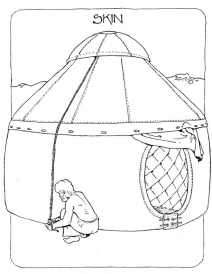

To stretch the skin tightly, push the bottom of each rafter up snugly against the canvas, as shown here.

THE STOVE

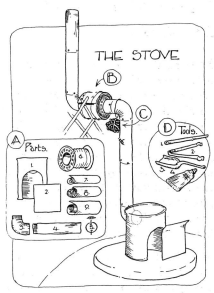

It's hard for me not to sound prejudiced, so let me say right out that I prefer the comforts of life and that even though I consider myself a nomad, my goal is always to be as cozy and protected from the elements as possible.

OTHER SIZES

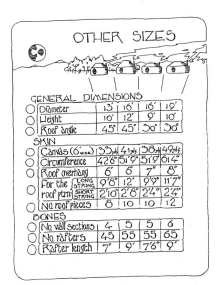

GENERAL DIMENSIONS

Diameter	13'	16'	16'	19'
Height	10'	12'	9'	10'
Roof angle	45°	45°	30°	30°

SKIN

Canvas (6' wide)	33 yd	41½	38¾	49¾
Circumference	42'6"	51'9"	51'9"	61'4"
Roof overhang	6"	8"	7"	8"
For the roof ptrn. LONG STRING	9'8"	12'	9'9"	11'7"
SHORT STRING	2'10"	2'8"	2'4"	2'4"
No. roof pieces	8	10	10	12

BONES

No. wall sections	4	5	5	6
No. rafters	45	55	55	65
Rafter length	7'	9'	7'6"	9'

I've learned the hard way that the quality of the final structure can only be as good as the quality of the raw materials, so choose your poles carefully.

THE SWEATLODGE

LUXURY EXTRAS

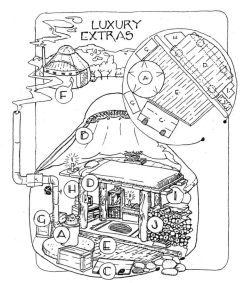

A: small woodstove / B: liner from floor to smokehole / C: fridge/root cellar / D: sleeping loft / E: multi-level floor / F: solar window

LESS is MORE
by d.price

I NEVER HAVE LIKED THE BUSY-BEE ADULT WORLD MUCH. AND AFTER SPENDING MOST OF 1989 STUDYING BOOKS ON THE LIFESTYLES OF NATIVE AMERICANS, I RETURNED TO MY CHILDHOOD HOME IN EASTERN OREGON WITH THE IDEA TO SOMEHOW LIVE IN THE MOST SIMPLE WAY IMAGINABLE. THE BOOKS HAD TAUGHT ME THINGS ABOUT THE EARTH AND US HUMANS THAT I HAD NEVER LEARNED IN ANY SCHOOLS. I REALIZED HOW UTTERLY SMALL I WAS IN RELATION TO NATURE AND THAT TO LIVE ON THE LAND IN HARMONY MEANT NOT EVER POURING CONCRETE ANYWHERE OR OVERUSING RESOURCES OF ANY KIND.

AS A KID I BUILT ENDLESS FORTS AND EVEN A SMALL CABIN. IN MY 20'S I DREAMED OF LIVING IN TIPIS. DURING MY 30'S I GREW TO RESENT LARGE RENT AND MORTGAGE PAYMENTS WHEN I REALIZED THAT THE SPACES I WAS PAYING ALL THAT MONEY FOR WERE NOTHING MORE THAN UGLY WOODEN BOXES. SQUARE, UNORGANIC COFFINS THAT IN NO WAY WHATSOEVER INTEGRATED WITH THE LANDSCAPE AROUND THEM. THEY WERE HARD TO HEAT IN THE WINTER AND COOL IN THE SUMMER. AND A FEW WERE EVEN FILLED WITH COCKROACHES!

BACK IN OREGON I BEGAN WORK ON SOME BIG QUESTIONS: (1) WHAT WOULD THE BARE ESSENTIALS BE FOR A LIVING SPACE? (2) COULD SOLAR POWER RUN THE NEEDS OF MY TINY BUSINESS? (3) WHAT DOES A HUMAN REALLY NEED TO LIVE IN A CLEAN, COMFORTABLE MANNER? (4) COMPLETELY IGNORING SO-CALLED "CONVENTIONAL WISDOM," AND STARTING WITH NOTHING, WHAT WOULD YOU NEED? (5) WHAT WOULD THE SHAPE, SIZE, AND BUILDING MATERIALS CONSIST OF? (6) COULD A PERSON'S HOME BE UTTERLY MINIMALISTIC AND EFFICIENT AND STILL LOOK AS ONE WITH THE LANDSCAPE?

FOR INSPIRATION IN THOSE BRAINSTORMING DAYS I OFTEN REFERRED TO THE *HANDMADE HOUSES* AND 1973 *SHELTER* BOOKS, WHERE I FOUND ENDLESS NEW IDEAS. I DREW PLAN AFTER PLAN, COMBINING SHELTERS FROM AFRICA WITH IRISH ROCK HUTS. I IMAGINED TIPIS OVER HIDDEN UNDERGROUND ROOMS. BUT IN THE END I DROVE TO THE WOODS TO CUT POLES AND BOUGHT A USED TIPI COVER AND SET IT ALL UP IN A RENTED MEADOW.

SPENDING THE NEXT FEW YEARS IN THAT MAGNIFICENT CATHEDRAL OF LIGHT BROUGHT ME CLOSER TO THE LAND AND THE WEATHER THAN I HAD EVER BEEN BEFORE. I RID MYSELF OF ALL UNNECESSARY POSSESSIONS AND LEARNED FIRST-HAND ALL ABOUT THE OLD ADAGE "LESS IS MORE." TO RUN MY MICROMAGAZINE BUSINESS I INSTALLED UNDERGROUND ELECTRICITY FOR MY OFTEN-USED COPY MACHINE. SOLAR WAS TOO SPENDY. NOW MY ELECTRIC BILLS RUN ABOUT $10 PER MONTH. IN THOSE EARLY YEARS I REMEMBER BEING ASTOUNDED AT HOW LETTING GO OF THINGS I THOUGHT I

Our kids Shane and Shilo spent two delightful summers in the tipi. During the winters i was kept busy shoveling snow off the tipi and trails. One winter the temperature went to -24°F.

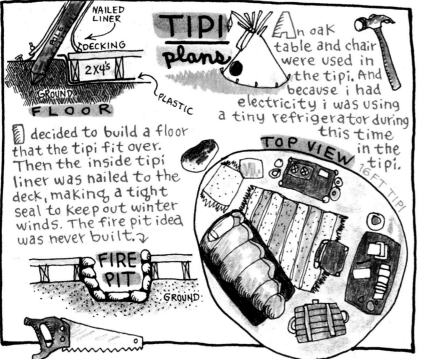

TIPI plans

NAILED LINER
DECKING
2X4's
GROUND FLOOR
PLASTIC

An oak table and chair were used in the tipi. And because i had electricity i was using a tiny refrigerator during this time in the tipi.

I decided to build a floor that the tipi fit over. Then the inside tipi liner was nailed to the deck, making a tight seal to keep out winter winds. The fire pit idea was never built.

TOP VIEW
16FT TIPI

FIRE PIT
GROUND

HUT plans

First i built a round deck with ½" pine boards. Holes were drilled 1ft apart all along the outer edge. Sturdy red willows were then stuck in the holes and bent to the center to begin forming the inverted basket shape. Exterior coverings were then added to the frame.

FRAME

FLOOR

PROPANE HOT PLATE WITH BOTTLE ON THE OUTSIDE.

BURLAP

SHINGLED VINYL WINDOW

H2O
HOT PLATE
FOOD
SUPPLIES
CLOTHES
COPIER
ROLLED UP BED FOR BACK REST
9X12 FLOOR PLAN

That's Shilo, above, reading at the tent site. The tent was very weather tight and warm and created an inviting glow inside on sunny days.

TOP VIEW

6X8 including vestibule

TENT plans

COULDN'T LIVE WITHOUT GAVE ME A HUGE BURST OF ENERGIZED FREEDOM. THE TIPI SERVED AS A WONDERFUL HOME AND TAUGHT ME MANY VALUABLE LESSONS FOR THE NEW STRUCTURES TO COME.

YURTS AND DOMES CAUGHT MY ATTENTION NEXT. I LIKED THE ROUNDNESS AND PORTABILITY OF YURTS BUT WAS DISCOURAGED BY THEIR PRICETAGS. UPON FURTHER RESEARCH I DECIDED THE MASS-PRODUCED AMERICAN MODELS LOOKED TOO STRAIGHT-EDGED WHEN COMPARED TO THE AUTHENTIC ORIGINALS FROM MONGOLIA. PUTTING PENCIL TO PAPER I DEVISED MY NEXT HOMEMADE HOME. MUCH LIKE A TRADITIONAL INDIAN SWEAT LODGE, ONLY LARGER AND WITH WINDOWS AND A WOODEN FLOOR. EUROPEANS CALL THEM "BENDERS." RED WILLOW SAPLINGS CREATE A WOVEN INVERTED BASKET SHAPE WHICH IS THEN COVERED WITH BLANKETS, PLASTIC AND THEN BURLAP.

THE HUT SERVED AS AN ULTRA-COZY, DRY, AND INEXPENSIVE STUDIO/HOME FOR THE NEXT TWO YEARS. I WOULD SPEND WHOLE DAYS DRAWING IN MY JOURNAL THEN RETURN TO THE HUT AND WRITE NOTES AROUND THE SKETCHES. I THOUGHT I HAD FOUND THE ULTIMATE SHELTER.

THEN MY LIFE TOOK A RADICAL TWIST AND I WAS OUT TRAVELING TO CREATE MY ILLUSTRATED JOURNAL UNDER THE GENEROUS SPONSORSHIP OF THE SIMPLE SHOE COMPANY. SINCE I DISLIKED MOTELS AND CAMPED OUT MOST OF THE TIME I DECIDED THAT WHEN I WAS BACK HOME I WOULD SIMPLY LIVE IN THE TENT! I DISMANTLED THE HUT AND ERECTED THE TENT IN THE SAME SPOT. STILL MY FAVORITE SHELTER, TENTS ARE NOW MADE TO WEATHER ALL FOUR SEASONS AND PROVIDE DRY AND EXCITING PLACES IN WHICH TO LIVE. THE ONLY DRAWBACK WAS THAT THE UNDERSIDE OF THE TENT FLOOR WOULD BEGIN TO GET MOLDY AFTER A WEEK SO I'D HAVE TO EMPTY OUT THE RUG, PAD, SLEEPING BAG, CLOTHES BAG, FOOD BOX, CERAMIC HEATER, BOOKS, LIGHT BULB AND CORD AND THE WATER BOTTLE FOR A GOOD ALL-AROUND HOUSE CLEANING!

more...

SHACK plans

It was great fun designing the smallest possible space for all my humble belongings and magazines. Even a tiny space below the rafters was used. After 2 years I had a thief break in and $5000 worth of cameras, computer and camping goods were stolen.

ADDITION 1999

6X10 ORIGINAL 1998
BUILT FROM RECYCLED WOOD

← HEAVY LOCKABLE DOOR

Eventually my magazine issues began to overflow the old Dodge's trunk so I decided to build a wooden house. A shingle-covered beach shack, with bed, table and book shelf. The 6x10 structure was so small I didn't even need a building permit. Total cost $95.

After living for years in the round and on the floor Japanese-style, the oak desk and chair felt odd and the room too boxy. I had a business to run though, phone messages to check, a website to build and valuables to keep locked up. The following year I added on a 6x5 bed/kitchen. A year later I followed my instincts of wanting to return to a round space and dug into the hillside and built an undergound 8 ft. kiva-like structure with an openable skylight for illumination. After some deep contemplations and having the shack robbed I decided to dismantle the wooden structure and scale down enough to just the kiva space.

With over a foot of earth on the roof the kiva is always at 50—60 degrees. To enter you must bow down and crawl through a short passageway, which is humbling and feels like what one ought to do when coming into a room that offers shelter and warmth from the elements. For fresh air I simply tip open the skylight and crack the door.

Many times I feel that this is by far the best home I've built in the meadow. Each item in my pared-down existence is immediately at hand and never lost! The thin bed mattress folds up

CUT-AWAY SWEAT LODGE

6 FT. CIRCLE

← WOODEN STRUCTURE WITH SKYLIGHT

PROPANE BURNER INTO LAVA ROCK FILLED STEEL PIPE

NEAR THE SWEAT LODGE IS A DOCK AND DRESSING ROOM

THINGS YOU WON'T FIND AT MY HOUSE	THINGS YOU WILL FIND AT MY HOUSE
1. Washer / Dryer	1. Lots of books
2. Hot water heater	2. Small radio
3. Sink or bathtub	3. Hot plate
4. Stove / Refrigerator	4. Simple foods
5. Furniture	5. Art supplies
6. Microwave	6. Sawdust toilet
7. Toilet	7. Camera
8. T.V.- Video- D.V.D.	8. Backpack
9. Stereo	9. Tents
10. Regular bed	10. Sleeping bag
11. Bed linens	11. Ceramic heater
12. Harsh cleaners	12. Biodegradeable soaps & cleaners
13. Vacuum	13. Water filter bottles →
14. Freezer	14. Hammock
15. Computer	15. Hand tools
16. Lawnchairs	16. Pushmower
17. Chainsaw	17. Bicycle
18. Lawnmower (gas)	18. Peace
19. Garbage cans	19. Quiet
20. Furnace/woodstove	20. Tranquility!
21. Sewing machine	
22. Car	
23. Mortgages!	

Dug into a hillside then completely covered with dirt, this tiny abode is toasty warm in winter and cool in the summer.

KIVA plans

KITCHEN
BED/CHAIR
BOOK SHELF
WALL TO WALL CARPET
← 2 ft passage

← TARPAPER WRAP (ants love to eat it!)
← HEAVY PLASTIC 2 LAYERS

5'
8 FT CIRCLE
BRICK FLOOR →

The only problem after 3 years is an invasion of big carpenter ants. They are eating the pine wood walls and coming in to gather around the radio antenna!

Cool Books

1. Handmade Houses
2. Shelter
3. Tiny Houses
4. Humanure
5. Circle Houses
6. Evasion
7. Walden
8. Being Nobody, Going Nowhere

Tools used to build all this

You'll notice the absence of a square or a level. I never got around to getting those!

d.price is the author of How To Make a Journal of Your Life, Moonlight Chronicles: A Wandering Artists Journal and the ongoing bi-monthly journal Moonlight Chronicles available at $5 each. You can order them from Dan at BOX 109, JOSEPH, OR. 97846
www.MoonlightChronicles.com

INTO AN EASY CHAIR. I'VE LEARNED TO READ AND WRITE IN THE EVENINGS INSTEAD OF STARING AT A TV. MY DIET HAS EVEN CHANGED TO THE POINT OF HAVING JUST ONE HOTPLATE AND NO REFRIGERATOR. LIVING IN THIS ALTERED MANNER CAUSES ME TO SPEND WAY MORE TIME OUT OF DOORS THAN WHEN I'M STAYING IN A CONVENTIONAL HOME. THE OUTHOUSE IS A SELF-COMPOSTING 5-GALLON BUCKET. (SEE THE BOOK HUMANURE). TO STAY CLEAN I TAKE DAILY SAUNAS IN A PROPANE-FIRED WOODEN SWEAT LODGE. THERE IS ALSO A BIG BUCKET ON THE ROOF OF THE DRESSING ROOM THAT CONTAINS A 110-VOLT WATER HEATING UNIT FOR AN OCCASIONAL GRAVITY-FED SHOWER.

LIVING IN THIS MANNER FEELS RIGHT TO ME. HAVING SPENT SOME TIME IN MY PARTNER'S HOME HELPING TO RAISE OUR TWO KIDS, I HAVE HAD THE LUXURY OF EXPERIENCING BOTH LIFESTYLES. IN THE MEADOW I DON'T USE UP SO MANY OF THE EARTH'S RESOURCES. AND BECAUSE I HAVEN'T BOUGHT IN TO THE IDEA OF MORTGAGES I CAN LIVE RELATIVELY FREE. FREE TO SLEEP IN IF I CHOOSE, READ BOOKS, WATCH NATURE AND LEARN TO CALM MY CONSUMER DESIRES, LIKE THE GOOD BUDDHA SAID!

I SOMETIMES WONDER WHERE MODERN MAN'S CONCEPT OF A LIFESTYLE HAS GONE. PEOPLE SEEM TO BE SO OVERWHELMED WITH TRYING TO KEEP EVERYTHING IN THEIR LIVES AFLOAT. MAYBE THE ANSWER TO THEIR DILEMMA LIES IN THE VERY WAY THEY ARE LIVING. IN THE BUILDING THEY CALL HOME. TO ME A HOME SHOULDN'T BE A BURDEN BUT ONE OF LIFE'S ULTIMATE JOYS.

—D. PRICE

NATIVE AMERICAN SHELTER

NAMES IN GENERAL AREA FOR REFERENCE. INCOMPLETE AND CONFLICTING RECORDS PROHIBITS TOTAL ACCURACY.

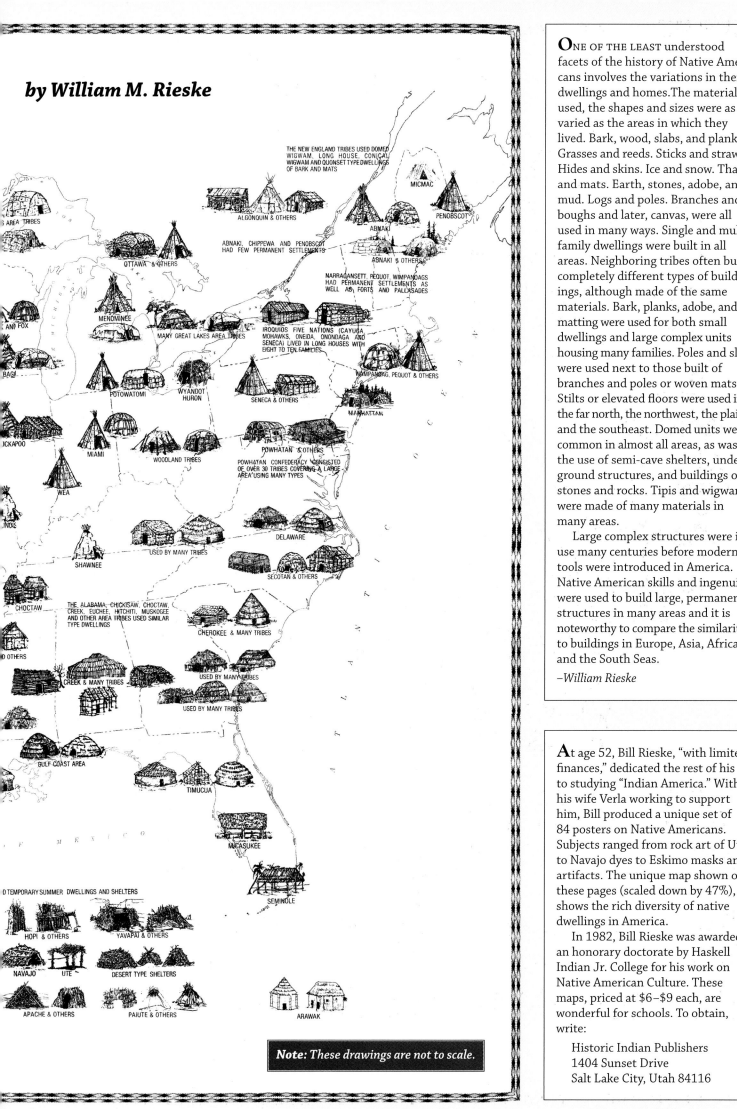

by William M. Rieske

THE NEW ENGLAND TRIBES USED DOMED WIGWAM, LONG HOUSE, CONICAL WIGWAM AND QUONSET TYPE DWELLINGS OF BARK AND MATS

MICMAC

...AREA TRIBES

ALGONQUIN & OTHERS

ABNAKI

PENOBSCOT

ABNAKI, CHIPPEWA AND PENOBSCOT HAD FEW PERMANENT SETTLEMENTS

ABNAKI & OTHERS

OTTAWA & OTHERS

NARRAGANSETT, PEQUOT, WIMPANOAGS HAD PERMANENT SETTLEMENTS AS WELL AS FORTS AND PALLASADES

MENOMINEE

...AND FOX

MANY GREAT LAKES AREA TRIBES

IROQUIOS FIVE NATIONS (CAYUGA, MOHAWKS, ONEIDA, ONONDAGA AND SENECA) LIVED IN LONG HOUSES WITH EIGHT TO TEN FAMILIES

...BAGO

POTOWATOMI

WYANDOT HURON

WAMPANOAG, PEQUOT & OTHERS

...ICKAPOO

MIAMI

SENECA & OTHERS

MANHATTAN

WOODLAND TRIBES

POWHATAN & OTHERS

...WEA

POWHATAN CONFEDERACY CONSISTED OF OVER 30 TRIBES COVERING A LARGE AREA USING MANY TYPES

...INOLE

DELAWARE

USED BY MANY TRIBES

SHAWNEE

SECOTAN & OTHERS

...CHOCTAW

THE ALABAMA, CHICKISAW, CHOCTAW, CREEK, EUCHEE, HITCHITI, MUSKOGEE AND OTHER AREA TRIBES USED SIMILAR TYPE DWELLINGS

CHEROKEE & MANY TRIBES

...D OTHERS

CREEK & MANY TRIBES

USED BY MANY TRIBES

USED BY MANY TRIBES

GULF COAST AREA

TIMUCUA

MICASUKEE

...D TEMPORARY SUMMER DWELLINGS AND SHELTERS

SEMINOLE

HOPI & OTHERS

YAVAPAI & OTHERS

NAVAJO

UTE

DESERT TYPE SHELTERS

APACHE & OTHERS

PAIUTE & OTHERS

ARAWAK

Note: *These drawings are not to scale.*

ONE OF THE LEAST understood facets of the history of Native Americans involves the variations in their dwellings and homes. The materials used, the shapes and sizes were as varied as the areas in which they lived. Bark, wood, slabs, and planks. Grasses and reeds. Sticks and straw. Hides and skins. Ice and snow. Thatch and mats. Earth, stones, adobe, and mud. Logs and poles. Branches and boughs and later, canvas, were all used in many ways. Single and multi-family dwellings were built in all areas. Neighboring tribes often built completely different types of buildings, although made of the same materials. Bark, planks, adobe, and matting were used for both small dwellings and large complex units housing many families. Poles and slabs were used next to those built of branches and poles or woven mats. Stilts or elevated floors were used in the far north, the northwest, the plains, and the southeast. Domed units were common in almost all areas, as was the use of semi-cave shelters, underground structures, and buildings of stones and rocks. Tipis and wigwams were made of many materials in many areas.

Large complex structures were in use many centuries before modern tools were introduced in America. Native American skills and ingenuity were used to build large, permanent structures in many areas and it is noteworthy to compare the similarities to buildings in Europe, Asia, Africa, and the South Seas.

–William Rieske

At age 52, Bill Rieske, "with limited finances," dedicated the rest of his life to studying "Indian America." With his wife Verla working to support him, Bill produced a unique set of 84 posters on Native Americans. Subjects ranged from rock art of Utah to Navajo dyes to Eskimo masks and artifacts. The unique map shown on these pages (scaled down by 47%), shows the rich diversity of native dwellings in America.

In 1982, Bill Rieske was awarded an honorary doctorate by Haskell Indian Jr. College for his work on Native American Culture. These maps, priced at $6–$9 each, are wonderful for schools. To obtain, write:

Historic Indian Publishers
1404 Sunset Drive
Salt Lake City, Utah 84116

NATIVE AMERICAN BUILDERS

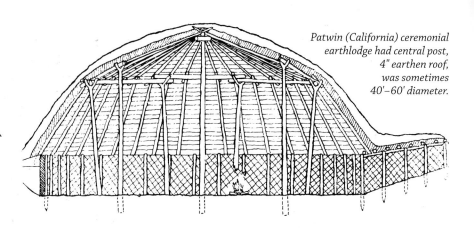

Patwin (California) ceremonial earthlodge had central post, 4" earthen roof, was sometimes 40'–60' diameter.

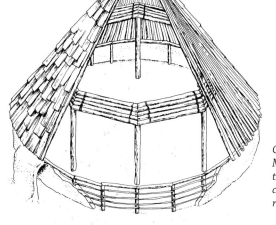

Creek (Alabama, Mississippi) town house with conical thatched roof 30' high

NATIVE AMERICAN ARCHITECTURE

PETER NABOKOV · ROBERT EASTON

Bob Easton and Peter Nabokov produced *Native American Architecture* (Oxford University Press) in 1989 and it remains the definitive work on pre–white-man building in North America. It shows the rich diversity of Indian buildings throughout the continent and covers ancient social customs, cosmological concepts, and ritual life as they affected building design. The book is lavishly illustrated, both with rare vintage photos, and Bob's wonderful pen-and-ink drawings, some of which are reproduced on these two pages.

Roof plank variations

Salish (Puget Sound, Washington) single-pitched shed house had 2 rows of posts, planks for walls and roof; in 1792 one was discovered near Seattle that was 380 yards long!

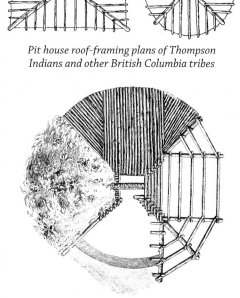

Pit house roof-framing plans of Thompson Indians and other British Columbia tribes

Delaware ceremonial Big House of logs and split shakes near Copan, Oklahoma, 40'× 25'× 18' (high)

Thompson Indian (southern British Columbia) pit house (above and below)

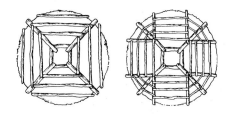

Iroquois (Lake Huron and upper New York State) longhouses were from 40'–400' long, 20'–30' wide

Conical hogan construction detail

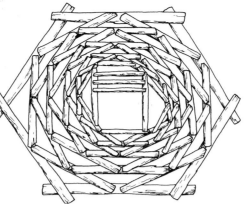

Navajo "whirling-log" hogan with cribbed-log walls and corbeled-log roof, was covered with 6" of tamped earth

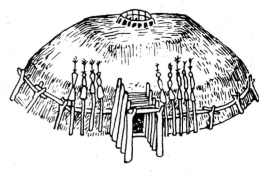

Hidatsa (North Dakota) earth lodge inhabited by Small Ankle, at Like-A-Fishhook Village, ND, in 1878

Conical forked-pole hogan (male)

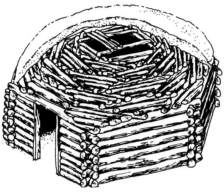

Corbeled log roof hogan (female)

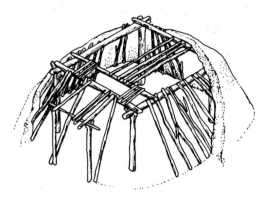

Four-sided leaning log hogan

Prehistoric southeast dwelling: poles set in trenches, walls covered with wattle and daub, thatched roof

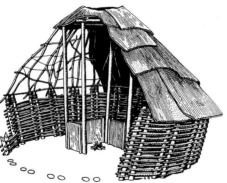

Adena (Ohio River Valley) circular house, outward-leaning poles, 20'–70' diameter

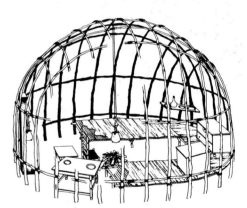

Kickapoo wigwam frame, 20' × 14' × 9' high, as built today in Nacamiento, Mexico and Eagle Pass, Texas

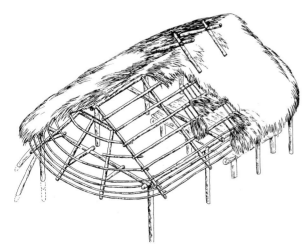

Chickee (southern Florida) house. Note diagonal bracing of corner posts at left. Palmetto thatch

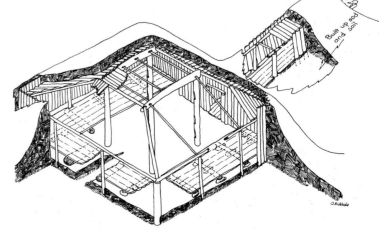

Mackenzie Delta Eskimo (Arctic) log-framed winter house with banked-up soil for insulation

197

THE TIPI: SERIOUS SHELTER

Robert Lewandowski

Of the three most critical necessities for life—food, water, and shelter—shelter is the most urgent and immediate need during the winter. A person can go without food for a month, water for a week—but an unsheltered night in a Wyoming blizzard could cause death by morning.

The elegant form of the tipi, perhaps the most beautiful of all Native American shelters, derives specifically from its function: to quickly provide a warm and comfortable living space, suitable for a stay of an entire winter if necessary, in areas of heavy snows, high winds, and sub-freezing temperatures. The tipi accomplishes this task by surrounding a campfire, and then providing the proper ventilation to remove the smoke. The fire provides warmth, light, and heat for cooking; the shimmering fire and glowing coals also furnish entertainment, creating an atmosphere suited to either quiet contemplation or animated storytelling.

The steep slope of the cone shape allows snow to slide off before becoming a crushing weight, and causes the wind to blow past the tipi forcing it down, but not over. The tipi is quick to pitch—starting with the sun on the horizon, the campfire will be cooking your food before the stars are out. At night, looking at it from a distance, the tipi lights up like a flickering Japanese lantern. Finally, at dawn, the whole camp disassembles and disappears like frost in the morning sunlight.

Thirty years ago, I spent two winters at 9000 feet on the Colorado Continental Divide in a tipi purchased from Nomadics Tipi Makers. Their thoroughly well-designed, handcrafted tipi allowed me to survive extreme conditions in comfort, both physical and spiritual.

Tipis have integrity.

Nomadics Tipi Makers
17671 Snow Creek Road
Bend, OR 97701
541-389-3980
www.tipi.com

BARNS

I LOVE BARNS. They're built for practical reasons (they have to work!), with economy, and attention to siting and weather. And, guess what? They're beautiful!

In a barn, you can see the framing — the posts, plates, braces, rafters, purlins . . . and they are invariably perfect. The architecture of economy.

Whenever I drive in the country, I'm on the lookout for barns. Usually they're deserted or at least no one's around and I go in and sit on the straw and admire the view, then shoot photos. They're my cathedrals.

Here are a few of the barns I've run across.

Barn lovers: We're working on a book on North American barns. If you can shoot photos of barns in your area, please contact us.

The barn shown in the three pictures on this page was owned by Randy and Luanne Queen. The roof is covered with hand-split shakes, and the siding is roughly milled cedar planking.

BARNS OF WASHINGTON

In 1973, MY SON Peter (then 12) and I headed north from San Francisco to catch the trans-Canada train to the East Coast. *(See pp. 146–151 — "Nova Scotia" — for the latter parts of this trip.)* We got on the train in Oakland, and disembarked in Seattle to take a detour to Spokane, where the 1974 World's Fair was being held. We got a room in a nearby boarding house, ate at hippy restaurants, and spent a few days at the fair. We then rented a car to drive back to the coast and catch the train for Vancouver. We headed for Seattle, west on Highway 2, and soon discovered it was a good choice: it was farm country, and there were barns along the route. (One of my favorite things to do is to cruise the countryside looking for unique farm buildings!) Farmers would see me shooting photos and tell me about other barns in the area.

When we got over to the main highway (Route 5), I decided to detour west to the Olympic Peninsula, through Sequim, and the picturesque town of Port Townsend. Photo-wise, this was a great choice, for this was dairy country, and there were beautiful barns along the (busy) highway.

The hayloft, with close-up of rafters, purlins, and hand-split roofing shakes

The rafters are 53'-long clear cedar poles, cut at a high altitude (where rings are tighter). It is 55'6" wide, 76' long. Posts are also poles (although the purlins are dimensional lumber), meaning much of this barn's materials was produced without a sawmill. You can tell the scale of the barn by Peter standing in the opening at the other end.

◄ *Previous page: Barn in countryside north of Toronto*

This large hay barn had an elegant, Japanese-style roof.

This unusual barn with poured concrete walls and graceful curved roof was abandoned.

A well-constructed log barn. The gambrel roof afforded the farmer more headroom in the loft (as opposed to a straight gable roof).

201

CALIFORNIA FARM BUILDINGS

THESE ARE photos I've shot over the years on numerous trips up and down the Northern California coast and countryside. These aren't the grand barns shown on other pages here, but rather minimal, graceful structures. They are ultra-lightweight due to no snow loads, fashioned by practicality and usefulness, and elegant as well as instructive in shape and construction.

Mendocino County

Mendocino County

Mendocino County

Mendocino-style, no-overhang shape that the Sea Ranch development unsuccessfully tried to copy in all its designs

Mendocino County

Sonoma County

202

This symmetrical barn is along Highway 5, north of Davis, California. Note the perfectly straight eave line, indicating no sag in the foundation.

Barn appears to be floating, near Boonville.

Horse barn with wings south of Big Sur

Mendocino County

203

ROUND BARNS

Octagonal log barn in Utah. Note the large cupola.

Octagonal barn off Highway 101 near Santa Maria, California

Round barn in Oregon. Nailed-together truss framing (two photos below) *radiates from silo in center.*

ROUND BARNS are rare. They are more difficult to build than rectilinear barns, and if subdivided inside they produce pie-shaped rooms. The most famous round barn in America is the magnificent stone Shaker barn at Hancock, Massachusetts. Eric Sloane, in *An Age of Barns*, says there was a saying that the round barn was designed "to keep the devil from hiding in the corners." Here are some round (and octagonal) wooden barns in America.

Peter French's 100' round barn in Oregon high desert country

more…

COWBOY CATHEDRAL

DRIVING SOUTH through eastern Oregon, I stopped in La Grande one morning to shoot pictures of some small homes. An old man was out watering his lawn and we started talking. I told him I was interested in buildings, especially barns, and he said, "Well, then you should go see the round barn." He said it was a large structure, built 100 years ago, near the small town of Frenchglen in southeast Oregon. He brought out a picture of it from his house. It looked beautiful!

I headed south and several hours later came to the Malheur Wildlife Refuge and followed some small signs to the barn. When I first saw it, it looked surprisingly small, probably because it fit into its surroundings so perfectly. From a distance it appeared to be a small conical shape, almost hovering above the fields. When I got up to the barn I was awestruck. It was huge, and perfect, sited at the end of a flat meadow, just on the edge of a sloping-up hill. It was a thrill to stand inside under the circular geometric framing. The carpentry was top quality, and the space was inspiring.

I spent a couple of hours there (alone), studying the framing and shooting pictures. In a while a pickup truck pulled up and three young people came in. They were all horse riders, from the local area, and had come to see the barn. They knew the history of the building and we talked about its perfect condition, over a hundred years after being built. One of them (Mike) said, "It's just like they left it yesterday."

≈   ≈   ≈   ≈

In June, 1872, 23-year-old Peter French set out for Oregon from Sacramento, California with 1200 head of select shorthorn cattle, six Mexican vaqueros, and a Chinese cook. Crossing the Sacramento River at either Redding or Chico (it's speculated), they drove northward into the Catlow Valley in eastern Oregon. There, French met a prospector named Porter. Down on his luck, Porter sold his small herd of cattle to French. Along with the cattle went Porter's squatter's rights to the west side of Steens Mountain and his "P" brand.

20'-wide paddock used for working horses during cold winter months

Three local horseback riders visiting barn

Further explorations in the area led to the discovery of a rich valley to the north, where melting snow from the west side of Steens Mountain meandered 40 miles south to Malheur Lake, producing lush grass — the Blitzen Valley.

French's operation expanded rapidly, as he acquired more land, more cattle — and more horses. With backing from California cattleman Hugh Glen, the French-Glen Livestock Company was formed. Native hay was cut and stacked, fences were built, drainage and irrigation of the valley began, more vaqueros and cowboys were imported, and hundreds of wild horses were captured and broken for freight teams, haying, and — buckerooing.

At its height, French's ranching empire encompassed 200,000 acres and 45,000 head of cattle, one of the mightiest cattle empires west of the Rockies. In the late '70s or early '80s, French built three round barns for breaking horses in winter months. The one shown here is the last, and it's a magnificent building. One hundred feet in diameter, the conical roof is framed with a 35-foot center pole of juniper (about 40″ at bottom, tapering to maybe 28″ at top), 14 surrounding juniper posts, and then a third wall of posts at the perimeter, about 8′ high.

French married Emma, the beautiful daughter of his partner Hugh Glenn. French built her a large, well-furnished "white house," with spectacular views, and the Donner and Blitzen River flowing past the front door. Yet Emma, who was said to be "flirtatious and worldly," left French for the bright lights of big-city San Francisco.

French was a little man with a big moustache, highly efficient in running his ranch and tough in dealing with homesteaders. The day after Christmas, 1897, French got into an fight with homesteader Ed Oliver. Oliver shot him in the head, and Peter French was dead at age 48.

Peter French was not only a skilled rancher. In reclaiming the valley's wetlands for pasture and haying, he also enhanced the habitat for migratory birds, which still arrive in profusion in the spring and fall.

Fisheye shows roof framing.

Note two different sections of framing.

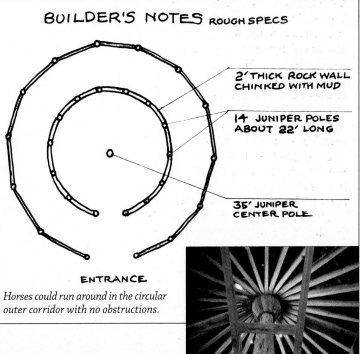

BUILDER'S NOTES ROUGH SPECS

2′ THICK ROCK WALL CHINKED WITH MUD

14 JUNIPER POLES ABOUT 22′ LONG

35′ JUNIPER CENTER POLE

ENTRANCE

Horses could run around in the circular outer corridor with no obstructions.

It all comes together here.

"[A cowboy's] . . . work started early each morning and lasted past the early dark. Riders had few chores to do at the home ranch; such things were for the cooks, a roustabout, or the newest man on the job. A wrangler arose at dawn and brought the horse herd into a corral. After breakfast the riders went out to pick their mounts for the day. Each one had a string of horses varying from eight to fifteen, depending on how many colts he was breaking. Experienced and regular riders always had to break out a bunch of young horses.

The individual rider decided what horse he wanted to ride and roped him, the choice depending on the work to be done that day and the qualities of the horse. If he were to be branding, he would need a horse used to roping; if he were going to ride with other men, he could ride and train a half-broken horse; if he were going to patrol a section of the country alone, he could use an older horse, past its peak for herd work — and safer. He led his chosen mount to the rack where saddles were kept, or to where he had left his saddle along a fence. If he were doing hard riding, he would need another horse by noon.

After Peter French had been in Oregon a few years, he had as good a string of cow horses as existed anywhere. As with cattle, he bred the cow horses up until they were fitted to the job. Old-timers who rode them described the P Ranch saddle band as good-sized for cow horses, with fine life and both strength and durability. Purebred stallions gave the size and speed, and the native cayuse provided the endurance and orneriness. Many of them needed to be broken all over again every morning, when a little bucking was expected by range riders; in fact, a little bucking was the mark of a horse's readiness for the day's work. Normally a cayuse never pitched long, just a few jumps to see, perhaps, if the rider himself were ready for the day's work."

Cattle Country of Peter French by Giles French, 1964

Photo: Joe Van Wormer

KEEPING THE TRADE ALIVE

In spring, 2003, I made one last photo trip for this book and drove up to the countryside near Eugene, Oregon to photograph the cob house of Ianto Evans and Linda Smiley (p. 84–85). On the way, I stopped off to see my friends Bill and Judy Pearl near Medford. As we drove down one of the backroads near their house to have breakfast at a cafe, I noticed a barn under construction; it looked just right. Judy said the builder had built a number of barns in the Medford/Ashland area, all of them timber frames, all mortised and tenoned together. Hmmm — sounded interesting, so after breakfast I made my way back to the barn and met builder Christoph Büchler, who was nailing up siding.

"Is it all pegged together?" I asked. Yes, he answered, the entire structure had been assembled with wooden pegs, no nails or bolts in the framework. Every post and beam was marked,

with Roman numerals chiseled in during pre-assembly in Christoph's yard, then used to erect the frame in place (on land owned by Joe and Mary Ellen De Luca). (See the two photos at left, photo at right, and one below it.) The numbering system could also be used for moving and reassembling the barn in the future, should it become necessary. The barn was 36′ by 36′.

I didn't see any power tools lying around and it turned out that all framing and sheathing was done with hand tools (no Skilsaws or nail guns). In addition, all the lumber was from the immediate area and consisted mostly of fir trees that were dying from either beetles or drought. It had been cut with a bandsaw mill in the woods, so a minimum of fuel was used in getting the lumber from the tree to the building site. This was unique in 2003!

While Christoph worked on the siding, two men were applying metal roofing (with drill guns). It was a beautiful building, it smelled good, felt right, used local materials, and was tuned into the environment.

I asked Christoph about his background. "Logging," he said. "I've always worked with wood. Furniture, buildings, you name it — wood." He started building mortise and tenon barns in the Medford area in 1983 and this was his tenth.

When I asked why no power tools, he said that the simplicity of it all appealed to him, that he didn't have to depend upon electricity and further, that he was also doing this ". . . to keep the trade alive."

GAMBREL BARN

THESE PLANS are for a 24-by-32-foot gambrel-roof barn of simple construction. Short lengths of lumber can be used and no large or heavy timbers are required. The haymow of this barn has a capacity of 15 tons of loose hay. Drawings from *Fundamentals of Carpentry — Volume 2, Third Edition*, by Walter E. Durbahn & Elmer Sundberg. Reprinted by permission of American Technical Society.

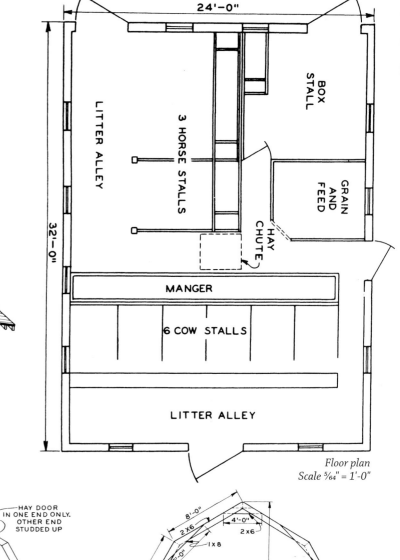

Floor plan
Scale ⁵⁄₆₄" = 1'-0"

Floor plan labels: 24'-0", 32'-0", BOX STALL, LITTER ALLEY, 3 HORSE STALLS, GRAIN AND FEED, HAY CHUTE, MANGER, 6 COW STALLS, LITTER ALLEY

Framing
Scale ⁵⁄₃₂" = 1'-0"

End-wall framing

Cross-section

Previous two pages:

Left: John Welles' mandala-like photo of timber-frame horse barn in western Connecticut. It is huge! 431 feet long, 56,000 square feet. All timber is fir. Built by Benson Wood Homes, Walpole, NH.

Right: Nailed-together barn in the Sacramento Valley near Winters, California, unfortunately going the way of many old barns.

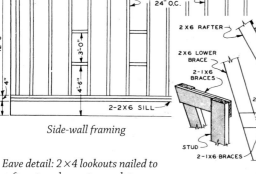

Side-wall framing

Eave detail: 2×4 lookouts nailed to rafter at angle great enough to carry roof water away from sides of barn

Eave detail

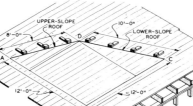

Above: Rafter bents are laid out on floor of haymow.

Point A = peak of roof. Note that the 8'0" and 10'0" rafter lengths are also shown on cross-section. The upper and lower rafter bents are held in place on floor with 2×4 blocks. Note how each rafter forms the third side of a right triangle.

OLD BUILDINGS

STONE STRUCTURES OF NORTHERN ITALY

Werner R. Blaser

WERNER BLASER is an architect, photographer, writer, and author of numerous books on architecture, engineering, and related subjects. In 1977, he published *Der fels ist mein Haus = The Rock Is My Home*, a tri-lingual paperback book of exquisite black and white photos and drawings of stone structures in Italy, Switzerland, and Ireland. Here are three photos of stone buildings in Northern Italy and a few quotes from this unique book.

What is so worthy of imitation in these secular stone buildings is the harmony of the interior, the load-bearing structure, and house corpus. All the dimensions are scaled to man, so that man and space form a unity.

Here . . . are people . . . who eke out an existence high above the vegetation line and have no alternative but to use stone for their houses . . . Everywhere there is unity between the architecture and the stone.

. . . on Alp Selva and San Romero we find the beehive-like drystone trulli, *which are round corbel dome structures built over a spring to serve as milk cellars. The corbel domes are constructed of ring-shaped layers of stone which diminish in diameter until the vault is closed.*

215

Buddhist shrine near Dingboche (14,107'), Khumbu (Everest) region

View east toward Tibet from Pheriche (14,200'), Khumbu

NEPAL-EVEREST 1996

Jim Macey

DURING SUMMERS I have worked as a "packer," packing mules in the Sierra Nevada mountains. In 1996 our pack station crew was invited by a Nepalese mountain guide to a take a pack trip in the high Himalayan mountains of Nepal and Tibet. In the fall of 1996, under the guidance of Jagat Man Lama, we spent 10 days traversing the Tibetan plateau, returning to Kathmandu, Nepal, where we flew in a Russian helicopter to Lukla (9200') in the Everest region. From Lukla we packed to the base region of Mt. Everest. Our gear was packed on yak/cow hybrids while we walked.

Yak/cow (zopjo) pack animal returning from Lobuche (16,164'), Khumbu

Buddhist prayer stones, Tengboche Monastery (12,369'), Khumbu

Pack stock returning from Mt. Everest

Along path from Tengboche Monastery on approach to Mt. Everest

At left and three photos above: Summer herders' cabins and stone corrals near the foot of Khumbu Glacier, elevation: 15,000'

Two photos above: Summer herders' cabins and stone corrals near the foot of Khumbu Glacier (15,000')

Swiss-built suspension bridge on the trail to Namche Bazaar, a Nepalese/Tibetan trading center (all freight transported by yaks) near the Tibetan border

Pastures at Dingboche (14,107')

View toward lowlands along trail into high Himalaya, above Tengboche, Khumbu region

Prayer (water) wheel on the trail from Lukla (9200') to Namche Bazaar

DISCOVERING TIMBER-FRAMED BUILDINGS

Richard Harris, A.R.A.

Central open trusses of medieval halls

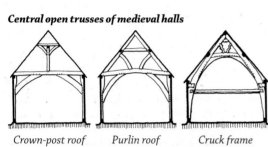
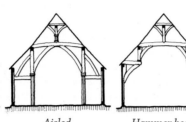

Crown-post roof   Purlin roof   Cruck frame   Aisled   Hammer beam

A medieval open hall

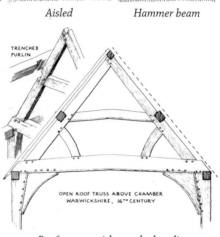

TRENCHED PURLIN

OPEN ROOF TRUSS ABOVE CHAMBER
WARWICKSHIRE, 16ᵀᴴ CENTURY

Roof trusses with trenched purlin

ROOF FRAME

FLOOR FRAME

CROSS FRAME

WALL FRAME

BAY BAY BAY BAY

Timber-framed buildings: bays and frames

ASSEMBLY   REARING

IN THE EARLY '70s, I met Richard Harris at the Architectural Association in London — a unique architectural school, with teachers and students from all over the world. At the AA, architecture was explored and discussed in all its varied forms, from space-age fantasies to the vernacular.

Richard was at that time a graduate student, doing beautiful drawings of 16th–18th century houses, cottages, and barns. In 1978, Shire Publications Ltd., a publisher of books about English building, crafts, and farming (www.shirebooks.co.uk), published Richard's *Discovering Timber-Framed Buildings*, a small book illustrated with these wonderful pen-and-ink drawings. These days Richard is the Director of the unique Weald and Downland Open Air Museum of Historic Buildings in West Sussex, England (www.wealddown.co.uk). If you are ever in London and interested in building, it's worth the trip to see these structures. *(See pp. 22–23 of Shelter for photos of buildings from the Weald Museum.)*

The purpose of this book is to show how beams were put together to form buildings. Buildings — at least those which survive today — were not home-made but were produced by carpenters who had served a long apprenticeship to learn the skills of their craft. Creating a building from trees is a bit like alchemy. Instead of turning base metal to gold, the alchemist-carpenter had to turn trees into beams, into frames, into buildings.

The secret of this magic was the craft tradition. This gave the carpenter a series of clear steps by which he could find his way through the maze of difficulties he faced in each new building. Each step gave a key to part of the process. Any change in these essential steps would upset the balance of his craft, but around them he was able to create the unique character of each building.

–Discovering Timber-Framed Buildings

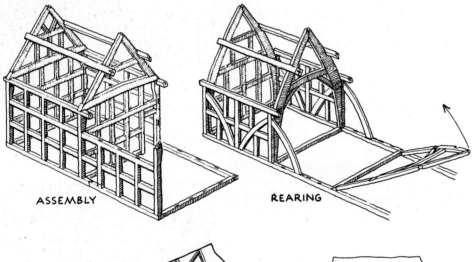

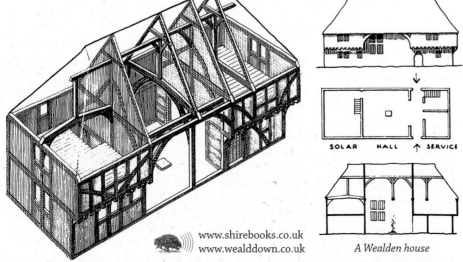

SOLAR   HALL ↑ SERVICE

www.shirebooks.co.uk
www.wealddown.co.uk

A Wealden house

Horse-powered grain mill from Vámosoroszi. Horses drive the huge wooden cogwheel around a central shaft which propels the mill stones. Originally built about 1800.

HUNGARIAN OPEN-AIR MUSEUM

IN 1991 I went to Budapest to meet with two of our authors. A travel agent I had contacted before the trip knew of my interest in buildings and offered to drive me to the nearby Szentendre Open Air Museum. The museum opened in 1974 and contains some 80 houses and 200 farm buildings from different regions of Hungary. It was a quiet afternoon, hardly anyone there and as we wandered through the streets it seemed like we had stepped back into the 18th century.

Plan of mill.
Stones at center, bottom

Cowshed and cart shed from Kispalád, northeast Hungary, mid-19th century

A house from Rédics, western Transdanubia, mid-19th century

A folding table and seat

A woodshed for firewood from Kispalád, northeast Hungary, mid-19th century

A supporting post in the barn from Kispalád, northeast Hungary, mid-19th century

A house from Vöckönd, western Transdanubia, mid-19th century

Tie-beam roof

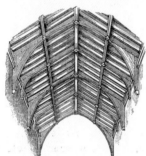
Trussed-rafter roof

Hammer-beam roof

Collar-braced roof

THE
OPEN TIMBER ROOFS
OF
THE MIDDLE AGES.
ILLUSTRATED BY PERSPECTIVE
AND WORKING DRAWINGS OF SOME OF THE
BEST VARIETIES
OF
Church Roofs;

BY

RAPHAEL AND J. ARTHUR BRANDON,
Architects,

Description of Terms

Bay. — The space between two trusses.

Braces. — Curved pieces of timber tenoned into the main timbers of the roof, and serving to stiffen and tie them together.

Collar-beam. — An horizontal piece of timber placed high up in the truss, and serving the double purpose of a stiffener to the principals and a tie to prevent their spreading outwards.

Hammer-beam. — An horizontal piece of timber lying on the wall-plates, at right angles with the wall into which the principal rafter and strut are tenoned.

King-post. — The strut which rests on the collar-beam, and into which the upper ends of the principals are sometimes framed.

Tie-beam. — An horizontal piece of timber extending from wall to wall, into which the ends of the principals were framed, and which served, as its name indicates, as a tie both to them and the side walls.

Little Welnetham Church, Suffolk

The drawings and text on this page are from The Open Timber Roofs of The Middle Ages, *by Raphael and J. Arthur Brandon, published by David Bogue, London in 1849. There are lovingly rendered, detailed pen-and-ink drawings of the roofs of various small churches throughout England, the majority of them in Norfolk.*

Accompanying each illustration in the book are construction details, including roof span, dimensions of tie-beams and rafters, spaces between trusses, etc. The book was reprinted in 1999 by Algrove Publishing Limited, Ottawa, Ontario, Canada.

Stowe Bardolf Church, Norfolk
Span of roof = 23'10"

HE SIMPLEST and earliest description of Roof was, doubtless, that formed by two rafters pitching against each other; it must, however, have soon become apparent that this mode was open to a serious objection, namely, that the rafters had a tendency to spread and thrust outwards the walls on which they rested; this led to the introduction of the tie-beam, which, in conjunction with the rafters, gives us that simple form of Roof which has been handed down to us in the earliest records we possess of any coverings to Buildings, and which, with some modifications, is still in very general use amongst us; and it must be admitted, that, in those cases where the Roof is intercepted from view, as, for instance, when concealed by a vault, no better construction can be adopted.

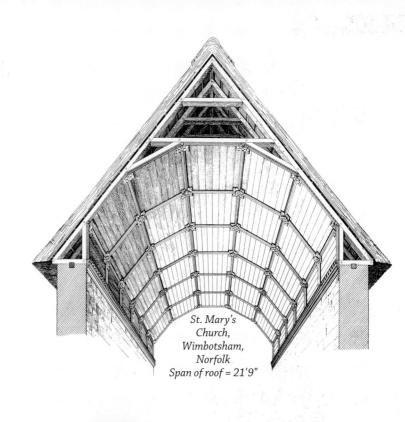

St. Mary's Church, Wimbotsham, Norfolk
Span of roof = 21'9"

Starston Church, Norfolk
Span of roof = 21'10"

NDEED MUCH has been done, much written and said on this interesting subject; the good that has been wrought by the revival of a purer taste in Architecture, and a recurrence to better principles, is but the harbinger of the good yet to be achieved. Architecture has at length roused itself from its slumber; or, to speak more correctly, has burst from the thraldom in which the vitiated taste of Puritanism had held it; it has risen, Phoenix-like from its ashes, to accomplish once more that beauty and perfection which made our Churches such worthy "monuments of love divine," such glorious works of fine intelligence.

Amidst the many beauties that these Sacred Edifices present to the admirers of Medieval Architecture, none are more striking than the taste and skill exhibited in the formation of the Roofs; and, indeed, there is no portion of a building, whether Ecclesiastical or Secular, requiring more skill in its construction, or that is more susceptible of ornament and decoration. Many of our Churches and Ancient Halls still attest the truth of this opinion by the evidence they afford of the matchless skill of the carpenter's art.

Limpenhoe Church, Norfolk
Span of roof = 17'0"

Porch of Neckington Church, Lincolnshire
Span of roof = 10'9"

WILLIAM COOPER, Limited,
Horticultural Providers

WILLIAM COOPER, LIMITED

IN THE '70s in an obscure little walk-up used bookshop in London I found a treasure: a little red book with gold and black lettering on the cover: an illustrated catalogue of one William Cooper, Ltd., of ". . . Horticultural Buildings, Garden Frames, Poultry Appliances, Rustic Work, Iron Buildings, Heating Apparatus & c. . . ." There is no date, but I'd guess it was turn-of-the-century.

The Cooper company manufactured portable greenhouses and plant frames, as well as chicken coops, duck houses, rabbit hutches, rustic furniture, and a variety of full-size buildings which were pre-cut at their London plant and shipped to customers at home and in the colonies. Here are a few of the greenhouse and plant frame (for starting seedlings) designs that seem as relevant today as they were 100 years ago. We plan to reprint the entire book in the near future.

Out of olde feldes, as men sayeth,
Cometh all this neue corn fro yer to yere,
And out of olde bokes, in good feyth,
Cometh al this newe science that men lere.
—Geoffrey Chaucer

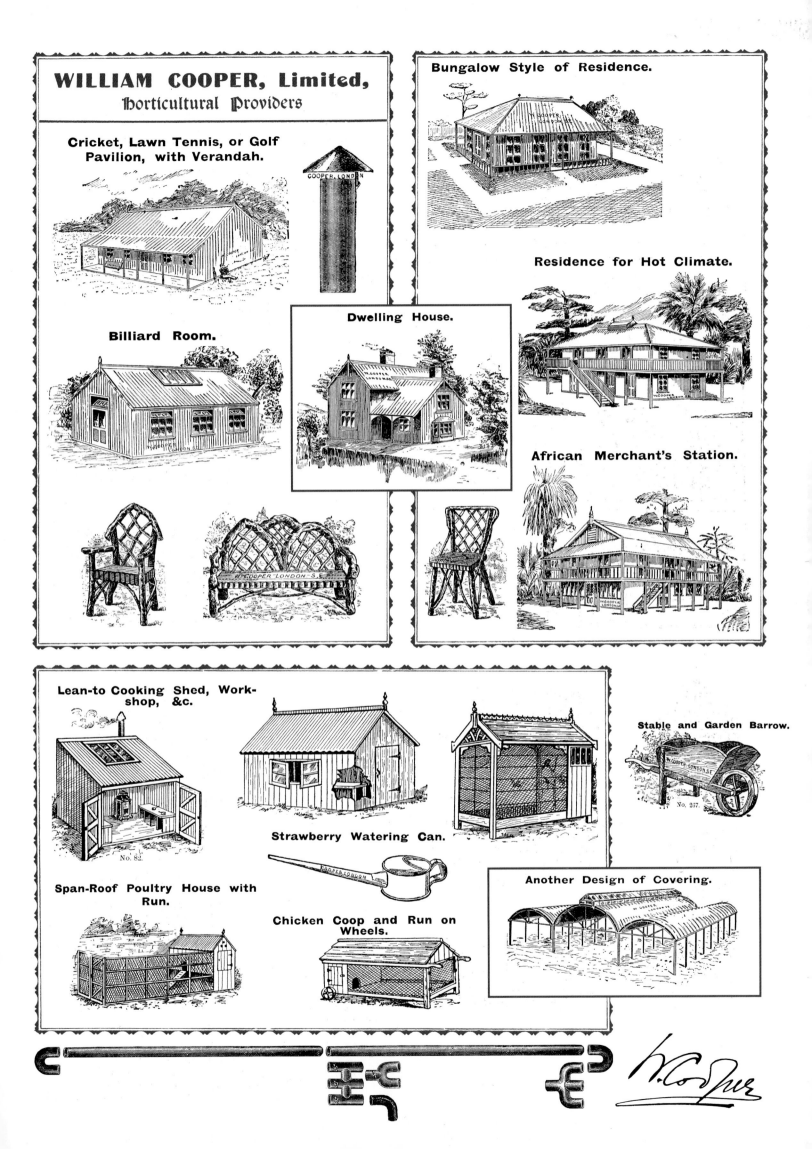

WILLIAM COOPER, Limited,
Horticultural Providers

Cricket, Lawn Tennis, or Golf Pavilion, with Verandah.

Billiard Room.

Dwelling House.

Bungalow Style of Residence.

Residence for Hot Climate.

African Merchant's Station.

Lean-to Cooking Shed, Workshop, &c.

No. 82.

Strawberry Watering Can.

Stable and Garden Barrow.

No. 267.

Span-Roof Poultry House with Run.

Chicken Coop and Run on Wheels.

Another Design of Covering.

BUILDINGS OF THE OLD WEST

Jim Macey

MOREOVER...

WE'VE BEEN WORKING on this book for a year and a half now and we're approaching the end. What's frustrating is that we have a lot of great material left over (actually, the makings of another book). Today I just ran through our files of unused material and pulled these few things out, shown on the following five pages.

Please note: If you have any material to contribute to the next book in this series — photos, drawings, stories, advice, building adventures, insights, excitement — please get in touch with us.

Below: Charcoal ovens near Ely, Nevada, built by Italian stone masons, called carbonari, *in 1872. They are the same general dry stone (no mortar) construction as the* trulli *of southeastern Italy, which are said to date back 5000 years. They were used to produce the fuel to run the smelters and mills in nearby Ward, Nevada. Although the doors are now missing, the stonework is in perfect condition, 130 years after being built.*

Brian Sell's wooden yurt near Baker, Nevada

I guess I have been building shelters all my life. I built tree houses and cave dwellings and huts from whatever was available. From 1968 to 1980 I owned a farm in the Sierras of central California. Recycling was very important to me during those years and I built my house and barn for less than $2000 using discarded and salvaged materials. In 1980 I moved to the island of Rarotonga in the South Pacific where there is little to recycle or salvage, but plenty of sand and rock. I decided to build a new shelter for my growing family which would be as maintenance-free as possible; a real challenge in the tropical climate near the ocean. Rainfall exceeds 130" per year, salt corrodes, hurricanes blow hard, and fire destroys. So I built my latest shelter of concrete. It will not burn, rust, wash away, or corrode. Building a concrete house is inexpensive, it lasts forever, requires little maintenance, but is very hard work, as we did it all by hand.

–Richard Wachter, Rarotonga, Cook Islands

. . . I'm enclosing a picture of an unusual hunting lodge (shelter) from northern Sweden — Lapland. This house was used by hunters who needed a secure place to sleep away from the wolves.

–Joel Lundberg,
Göteborg, Sweden

Northern California

Window detail, Santa Cruz, California

Interior of ferro-cement dome in woods near Belmont, New York

Bedrooms

We all dwell in a house of one room—the world with firmament for
its roof—and are sailing the celestial spaces without leaving a track.
                                                        —John Muir

more...

Southern Wyoming

Near Cazadero, California

Tuscarora, Nevada

Poured concrete/rock walls, Tuscarora, Nevada

Towers

Talent, Oregon

Some town in Oregon

San Anselmo, California

El Pacífico

more...

Godfrey Stephens

In 1964, I met Canadian artist Godfrey Stephens in Yelapa (near Puerto Vallarta), Mexico. We had driven there from SFO in a '60 VW van. Godfrey had painted murals and was carving wooden sculptures in the little village and living in a $5-a-month palapa shack looking down on the blue Pacific Ocean. He was ready to head home to British Columbia, so returned to the States with us. We've kept in touch off and on through the years. He is a true artist and wild spirit (and sailor of the world's seas). Here's an email from him a few years ago.

Dear Lloyd . . . thanks for the letter, and the web stuff, and especially your cousin's miraculous cavern and fitted rock work floor. Where is that beach shack you sent with the first letter? . . . any way, when returning from Many years on the road, I went to a remote Wild Pacific beach, Wreck Bay (Florencia Bay) between Ucluelet and Tofino, west coast of Vancouver Island in what is now Pacific Rim National Park, it being a very Wild place, Like Calif. coast, but no people. Built a Plazarium, with carved yellow cedar, classic Haida-style pole holding the ridge pole up, ships' debris, flotsam & Jetsam, a 10-gallon oil drum with a hole chopped into it, and an old drain pipe from the Dump for a chimney, and polythene plastic, gets a good, almost hermetic seal, when it melts around the chimney pipe!, to be done when wet and raining. Well I did Lots of sculpture there, and eventually when they kicked me out of the place when creating the Pacific Rim Park, and running an asphalt road in and to me Ruining the place, Pay parking from the wardens office, you know the horrid march of Humanity to the edge. Ironically, the Park people got wind of the Wild Carver guy living the, to-me, perfect life, Lots of Free love coming around in the summer you know. Just a wonderful period, 1968 to 1971. The parks bought, or rather gave me the price of a Klepper Arieus sailing Kayak which folded down into two bags, for an abstract red cedar sculpture, which Jean Cretien presented to Princess Anne when she pulled the silk off the Bronze Placque Inaugurating the new "Pacific Rim Park" (which the publicity led to some Great future Commissions) and Lloyd you should have seen the incredible shacks the Freaks built along the beaches and Driftwood-choked Bays, any way, I later built a Herringbone shack 7'× 11' right on the outside of Wickaninnish Island, a MOST perfectly beautiful Beach, sometimes we (My Wild Girl and me) wouldn't see another Humanoid for months, we had a Daughter Tilikum, who is 26 now and just bought her first Sailing boat, a William

Atkin Eric Jr. in Port Angeles. She was born on the deck of my 34' gaff-rigged Wharram Catamaran, Tompatz, which I put over the distance of a circumnav of the world on, sailing mostly without the Bloody Seagull engine. Being Driven into the sea by the land people's progress, I finally got sea legs and have built quite a few boats, and have barely looked back since the Kayak.

You must be interested in sailing boats because of your Shelter books and the old collectors items, the Dome books 1 and 2 . . . My 36' steel sloop is buried under the sands off Punta Marquez, Baja del Sur. My second sail to Mexico on same boat lived aboard for 8 or 9 yrs. and came back, and went through a slew of different craft and finally at great difficulty wangled money from my Art, the only way I have ever worked, and got this fine Hull Ron Pearson built for himself, y'know another child, and the boat went on the back burner. Having spent time & Money on my Master welders labour converting and cutting all the things not liked about the unfinished hull, to be as best as can be from what we have to work with. It is 40' l.o.a.× 12' beam, and with the Twin Keels draws about 4'4". Steel pilot house African Queen canopy, so strongly built, it braces the main Gallows on which the Blocks are secured for side sheeting to a 900 square foot (rip stop U.V.-protected Tarp material 1200 sq.ft. for $300 cdn.!) fully Battened Asian LUG sail, on a 12" diameter Unstayed solid fir mast. with the twin Fins & Skeg can sit on a three point landing when the tide out, Bottom serviced without expensive weights, the Skeg is hollow and circulates an enclosed heat exchanger which has no salt water running thru the engine which is a 56 h.p. Isuzu Diesel (new) and

The Ching Ting "(Dragonfly" in Cantonese), a boat slammed together to get sea borne again after the wreck of the first Mungo

new Lead ballast ready to saw up and pass through the underwater window hole, to drop into the Incredibly built hollow keels and dump the lead "Saw Dust" on top and tiger torch it solid, drop a steel lid, weld tight, then no oxygen gets in to cause corrosion, the stantions are stainless steel like the Bull Cap and the Rails all around coast guard style Solid pipe NO wire, all Stainless recycled, i.e., the exhaust system is Stainless piping from Island Farms dairy, Milk used to course through those shiny pipes! Easily taken apart with big Nut Joiners this is a sail boat, but it is built with lovely lines for a steel hull, and tough as a working tug. In Cabo once aboard Mungo, there was the famous 40' double ender the "Joshua" which Bernard Motessier had sailed around the world 1½ times non-stop, etc. It was wrecked there but relaunched by friends from Port Townsend, they wanted to trade that

venerable Old ship for mine, I declined.

Oh Lloyd this is turning into a wild story . . . have a friend Bruno Atkey who builds Wonderful Houses out of driftwood and shakes, they are NOT hippy dippy but real elegant simple structures, The Indian Long House at Hesquiat Just in from Estevan Point (on chart) and the Buckland Farm complex resort at the old Cougar Annies, right on the beach! He would be a good guy for you to fotograph for a feature or a few inserts in another Book, hey and please put some Boats in, they are a lot better built than most houses and have Lines that fit the Most powerful element. Enough! Megan has filed a lot of my crazy history . . . gotta go to bed. Hope to resume contact with you old amigo, Yelapa was a long time ago . . .

–Godfrey Stephens
godfreystephens@shaw.ca
www.godfreystephens.com

Tlaook, Maryanna, Godfrey, and Cos in front of Godfrey's boat

BOOKS

BOOKS ON BUILDING

The Top Three Books on Our List

These three books: *Built By Hand, Dwellings,* and *Micro Architecture* are extraordinary and unique, as well as timely. They are in a class by themselves.

We HAVE A LIBRARY of over 800 books on building, collected over the last 40 years. Most of them are on homes and small buildings, handmade and owner-built. Here are reviews of some of the best. (I'm not reviewing those books already mentioned in other parts of *Home Work*.) A few things to note here:

1. This list isn't inclusive: there are plenty of other wonderful books on building—we can't cover them all.

2. I haven't covered books that are already well-known, but chosen those I think are outstanding, and that you may not know about.

3. A number of these are out of print, but with online searching, most of them can be found.

Built By Hand

Yoshio Komatsu; text by Bill Steen, Athena Steen, and Eiko Komatsu
2003; 472 pages; hard cover
$50.00
Gibbs Smith Publishers, Layton, Utah

There has never been a photographer of buildings like Yoshio Komatsu. He's been shooting photos around the world for 25 years now and this stunning book is the result. It's a spectacular photo journey, like a film, around the planet, giving you rare glimpses of shelter in every corner of the earth. The photos are absolutely wonderful. Text and editing are by straw bale gurus Bill and Athena Steen, and Eiko Komatsu. This is like an updated, expanded, technicolor, improved-upon *Architecture Without Architects.* Wow!

Micro Architecture

Edited by Kiyoko Semba and Kesaharu Imai
World Photo Press, Tokyo, Japan
2003; 478 pages; soft cover; $55.00
Available in U.S. from Ursus Books and Prints, Ltd., New York.
212-627-5370
ursus@ursusbooks.com

Every architect should own this book. There is no other book like it. It contains thousands of photos, as well as drawings, imaginative collages, and unique layout. It covers mostly small buildings—homes, barns, sheds, yurts, treehouses, tipis—and just about anything visual that caught the photographers' and editors' eyes. The layout is imaginative and stunning. The book makes the reader wonder how anyone could gather so much information, and then assemble it into a cohesive whole. In addition to architects, I'd recommend this book to builders, designers, artists, photographers, and anyone fascinated with the visual world. Unique and inspirational.

Dwellings
The Vernacular House World Wide

Paul Oliver
Phaidon Press Inc., NY
2003; 288 pages; hard cover
$59.95

Paul Oliver is a scholar with soul. In the early '70s, he published *Shelter and Society,* which was a major influence in our compilation of material for *Shelter.* He was formerly head of the Graduate School at the Architectural Association in London and is now Chair of the Master's Course in International Studies in Vernacular Architecture at Oxford Brookes University in Devon, England. *Dwellings* is about handmade buildings by indigenous people of the world, some of whom still thrive, others whose traditional ways of life are threatened. The photos, mostly by the author, are wonderful, and the accompanying text is perceptive and informative.

The American House

Mary Mix Foley; Drawings by Madelaine Thatcher
Harper Colophon Books, New York
1981; 300 pages; soft cover

Although out-of-print, this book is well worth tracking down. It covers over 300 American houses, from early American vernacular architecture imported from Europe, to Georgian, Greek, Victorian, and Beaux Arts styles, to Modern architecture. The pen-and-ink drawings are extraordinary. Here's what the *Chicago Tribune* said about it: "The most comprehensive, lucid, and best-illustrated guide to U.S. house styles ever published. It covers everything from 17th-century huts to the Inclusivist baloney of Venturi and Rauch, and belongs on everyone's ready-reference shelf."

Connecticut Saltbox

Cajun Cottage, Louisiana

French "Raised Cottage," New Orleans, Louisiana

French Vertical Log House, Mississippi Valley

Banua Toraja
Changing Patterns in Architecture and Symbolism Among the Sa'dan Toraja, Sulawesi, Indonesia

Jowa Imre Kis-Jovak, Hetty Nooy-Palm, Reimar Schefold, Ursula Schulz-Dornburg
Royal Tropical Institute, The Netherlands
1988; 136 pages; soft cover
$32.50

The Sa'dan Toraja are a tribe living in the highlands of South Celebes, Indonesia. Their soaring and elegant houses and rice barns were documented by a photographer, architect, and two anthropologists in 1983 and this striking book is the result. The buildings are on stilts and held together with pure joinery (no nails). They are often decorated from bottom to top, and the soaring roof forms with upswept ridgepoles are spectacular examples of vernacular

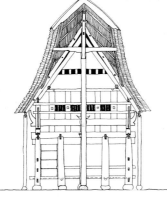

architecture. This book combines photographs and drawings of construction and technical details with descriptions of the social and religious significance of the Toraja buildings.

Dwelling house on polygonal piles in Nanggala district, with two carved buffalo heads

The Beauty of Straw Bale Houses

Athena and Bill Steen
Chelsea Green Publishing Company, White River Junction, Vermont
2000; 116 pages; soft cover
$22.95

This is a small book with beautiful color photos by the authors of the best-selling (120,000 copies) book, *The Straw Bale House (see pp. 74–81)*. It demonstrates what the authors have learned in their work using natural and local materials: that homes built this way can be creative and beautiful. As well, they feel good: ". . . the walls surrounding us day in and day out need to embrace us, our dreams and our passions woven into their very fabric."

more...

Early Mexican Houses
A Book of Photographs and Measured Drawings

G. Richard Garrison and George W. Rustay, with preface to the new edition by David Gebhard
Architectural Book Publishing Company, Inc., Stamford, Connecticut
1990; 174 pages; hardcover
$49.50

In the late 1920s, architectural draftsmen George Garrison and George Rustay set out with camera and pen to document select examples of "minor domestic architecture" in Mexico. Their book was published in 1930 and was more recently reprinted. It is entirely graphic (no text other than the foreword and preface) and contains black and white photos, along with wonderful architectural drawings showing perspective, elevations, floor plans and details. (The drawings are very much

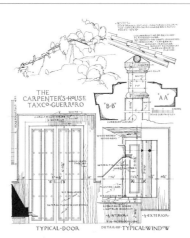

like those in a hand-lettered 1932 copy of *Architectural Graphic Standards* we have in our library.) The book was intended to be useful for architects as well as engaging for the general public.

The French Farmhouse

Elsie Burch Donald,
Illustrated by Csaba Pasztor
Little, Brown and Company, London
1995; 200 pages; hard cover

Describing the old farmhouses of different regions of France, this book is profusely illustrated with over 100 photos and some 300 line drawings. It explains how farmhouses were built and furnished and describes the life that went on within. The first half of the book covers materials, construction, and peasant life, while the second half covers farmhouses in 26 regions in France. Of interest to anyone interested in European folk building, and especially to those interested in buying a house in France.

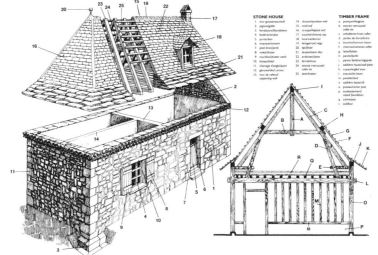

Houses by Mail
A Guide to Houses From Sears, Roebuck and Company

Katherine Cole Stevenson and
H. Ward Jandl
The Preservation Press, Washington, D.C.
1986; 366 pages; soft cover
$27.95

Between 1908 and 1940, Sears, Roebuck offered pre-fab houses from special catalogs. Over 100,000 of these were built in America. This heavily illustrated guide shows 447 different models, each with a rendering of the house, along with floor plans. There are a lot of books out there on early American house plans, but this is the most useful I've seen, a treasure trove of ideas for designing small homes. The floor plans are tiny (you'll need a magnifying glass), but it has allowed the authors to pack in a great deal of info.

The Houses of Mankind

Colin Duly
Thames and Hudson, London
1979; 96 pages; soft cover
$5.00–$10.95 used

A more descriptive title for this small book would be *The Tribal House*. In any event, it's a gem. The focus is on tribal domestic buildings, and on how their design and decoration are influenced by social customs and religion. It is somewhat like a small version of Paul Oliver's *Dwellings (see p. 234)*, with carefully chosen photos that will appeal to any lover of vernacular architecture, along with well-researched descriptions of the forces that produced the designs.

Dogon meeting house, Mali, Africa. Access is guarded by two female figures on posts facing each other.

Japanese Joinery
A Handbook for Joiners and Carpenters

Yasuo Nakahara; translated by Koichi Paul Nii
Hartley and Marks, Publishers, Vancouver, B.C.
1990; 240 pages; soft cover; $29.95
800-277-5887
pbdesk@hartleyandmarks.com

Japanese wood joinery is a master craft dating back to the 7th century. It is imbued with reverence for not only the carpenter's craft, but for "the spirit of the tree." In Japan, the beauty of wood in a building is considered as important as structural strength. This profusely illustrated book includes more than 100 splicing (*tsugite*) and connecting (*shiguchi*) joints, with details and step-by-step instructions for cutting and assembly. Both very simple and highly complex joints are included. In some cases modern simplifications of more difficult traditional joints are presented. The drawings are wonderfully clear and instructive.

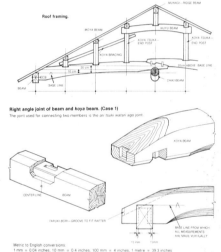

Japan's Folk Architecture
Traditional Thatched Farmhouses

Chuji Kawashima
Kodanshha International, Tokyo
1986; 260 pages; hard cover
$48.00

A comprehensive, graphically exquisite book of the *minka*, or traditional Japanese farmhouse. This book could be the prototype for any book on regional, vernacular architecture. There is a map of Japan, showing the different *minka* styles for each region of the islands and more than 400 clear black and white photos and accompanying pen-and-ink drawings. The building materials — earth, wood, and stone — come from the mountains and forests that surround the houses. The designs are said to have originated in Japan's prehistoric past. They vary from the steep thatched roofs of the north, with its heavy winter snows, to small low buildings in the south that have raised floors to maximize ventilation and minimize flood damage. The author is an architect who has spent over 50 years studying, drawing, and photographing Japanese farmhouses.

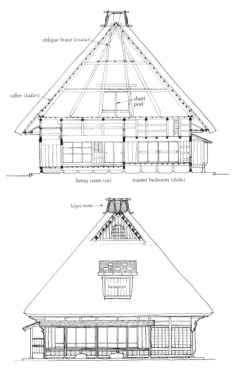

more...

Lehman's Non-Electric Catalog

Free from Lehman Hardware and Appliances, Inc., One Lehman Circle, Kidron, OH 44636
www.lehmans.com
888-438-5346

Hundreds if not thousands of items for home, garden, and farm. "We offer everything you need to live without reliance on electricity." A huge selection of kitchen tools (hand-cranked milkshake-making blender, cast-iron cookware, apple presses, food dryers, etc.), garden carts, composters, froes (for splitting shakes), water pumps, Bag Balm (for chapped hands as well as goat udders) and on and on. The catalog is a delight to leaf through, especially in this electronic age. It's not only essential for homesteaders and owner-builders, but can be useful to anyone living anywhere who is interested in doing (at least some) things the old way where the result is well worth the effort.

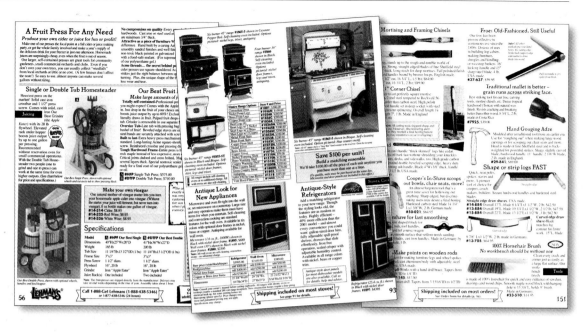

The Living House
An Anthropology of Architecture in South-East Asia

Roxana Waterson
Watson-Guptill Publications, New York
1990; 264 pages; soft cover
$16.00– $40.00 used

How people shape buildings and buildings shape people. This is a scholarly work by an anthropologist about indigenous building in South-East Asia. It examines ". . . ideas and beliefs about buildings which are regarded as powerful, sacred, or alive." It covers the role of kinship in house design and how social relations define the uses of spaces within the house. There are a lot of great photos, including a number of archival black and whites (some dating back to the turn of the century) that are unique.

Conical houses of Managgarai people, West Flores Island, Indonesia. Circular roofs reach to ground.

Minangkabau (matrilineal society) Islamic school, West Sumatra, Indonesia. Note both traditional thatch and galvanized metal roofing.

Lucarnes

Yves Brondel
Éditions H. Vial, 8, Rue des Moines, 91410, Dourdan, France
Phone: 011-33-1-64 59 70 48
Fax: 011-33-1-64 59 52 96
h.vial@wanadoo.fr
1993; 144 pages; soft cover

Lucarne is the French word for dormer, and this unique book with beautiful pen-and-ink drawings shows about 100 different dormers on French buildings. (A dormer is a window rising out of the roof and lighting the room inside.) Text is in French, but the drawings speak for themselves and are a great reference for both builders and architects. The dormers are divided into groups, the most typical being those with gable roof and flat front faces, but others shown have rounded, peaked, or geometric roofs, and there are a number of charming "eyebrow" dormers.

Primitive Architecture

Enrico Guidoni; translated by Robert Erich Wolf
Rizzoli International Publications, Inc., NY 1987; 224 pages; soft cover
$30–$120 used

A very wordy, scholarly book with over 100 illustrations of the architecture of societies that "... have remained outside the large, highly organized political entities into which the modern world is organized." There are examples of about 200 building techniques, with wonderful photographs. The author empasizes the role of architecture (building) in the social, cultural, and economic development of people in transition from hunter/gatherer existence to villages and finally to urban life. Great archival photos.

A Roof Cutter's Secrets
To Framing the Custom Home

Will Holladay
Journal of Light Construction 1989; 340 pages; soft cover
$32.50 (from Journal of Light Construction or Builders Booksource — see p. 240)

Will Holladay is a master carpenter specializing in framing. He has written (and illustrated) this clear and useful little book on framing a wood-frame house; walls, all aspects of roof framing (including cutting, hips, valleys, dormers, etc.) as well as great tips on framing circular towers and stairs, arches, bays, and skylights. The author intends this for the experienced carpenter, but it could also be useful for an owner-builder who wants to know how the pros work. Drawings and photos by the author; a fine and useful book.

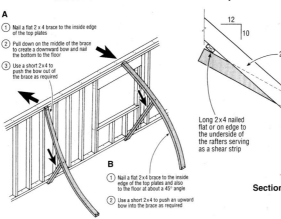

To Push or Pull a Wall

A
1. Nail a flat 2 x 4 brace to the inside edge of the top plates
2. Pull down on the middle of the brace to create a downward bow and nail the bottom to the floor
3. Use a short 2x4 to push the bow out of the brace as required

B
1. Nail a flat 2 x 4 brace to the inside edge of the top plates and also to the floor at about a 45° angle
2. Use a short 2x4 to push an upward bow into the brace as required

Split-Pitch Tail Kick-Up

2x10 common rafters
4x6 beam tails
Long 2x4 nailed flat or on edge to the underside of the rafters serving as a shear strip
Exterior wall

Section View

King-Pin Octagon Beam Tower

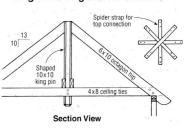

Spider strap for top connection
6x10 octagon hip
Shaped 10x10 king pin
4x8 ceiling ties

Section View

A Shelter Sketchbook
Timeless Building Solutions

John S. Taylor
Chelsea Green Publishing Company, White River Junction, Vermont 1997; 164 pages; soft cover
$18.95

This is a comprehensive little book, with over 600 simple pen-and-ink sketches by the author, an architect in New Hampshire. The focus is on indigenous buildings all over the world, in which "... the unselfconscious utilization of common sense can yield elegantly simple, practical, and timeless solutions to the most basic needs addressed by human shelters." It is divided into three sections, covering: the environment (sun, wind, cold, water); human needs (sleeping, cooking, eating, bathing); and the building itself (roof, walls, floor, doors, windows). The author's goal is to inspire builders to utilize the vast and rich wisdom of folk architecture in designing for today's world. An excellent first book for any student of architecture, and for anyone building their own shelter.

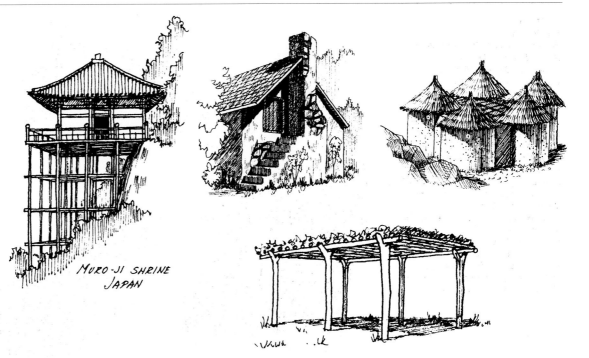

MURO-JI SHRINE JAPAN

more...

Traditional Chinese Residences

Wang Qijun
Foreign Language Press, 24
Baiwanzhuang Rd., Beijing, 10037, China
www.flp.com.en
2002; 108 pages; soft cover
$29.95

Books on Chinese architecture are rare (think, by contrast, of all the books on Japanese building and architecture); books on vernacular Chinese even rarer. This unique little book covers development of traditional Chinese residences from 7000 years ago (matriarchal societies) up to the Quin Dynasty (1644–1911). It covers a number of house designs from different provinces in China, including cave dwellings, earthen buildings, fortified villages, houses on stilts, and Mongolian yurts. The color photos are consistently good, and the residences shown are often unusual and not seen in other books on building. Text is minimal but informative.

Houses on the water, Zhouzhuang, Suzhou, Jiangsu Province

Tropical Bamboo

Marcelo Villegas
1990; 176 pages; hard cover
Rizzoli International Publications, Inc., NY;
a 3rd edition was published in 2001 by
Villegas Editores Ltda, Bogota, Colombia.
When we went to press, 3 used copies
were available from www.abe.com, for
$45–$65.

Illustrates the many uses for bamboo: country houses, bridges, corrals, water pipes, furniture and instruments. *Bambusa guadua* is a species of bamboo that grows in the Old Caldas region of Colombia; it is called by natives "the gift of the gods." Included is the innovative and elegant work of Colombia architect Simón Vélez. The color photos are excellent and the book is an inspiration for anyone working with bamboo. All photos were shot in Colombia. *Hey, Rizzoli editors: Reprint this book!*

Bambusa guadua

Orchid nursery in Caldas. Structural supports are called "chicken legs."

Hundred-year-old three-story bamboo coffee-drying plant on river bank in Caldas, Colombia. Note heavy tile roof.

Two Publishers

We recommend that you check out the websites and/or get catalogs from the following two publishers, both of whom publish many wonderful books on building and the related arts:

Gibbs Smith publishes books that are smart, stylish, and sophisticated, on a variety of subjects, including building and architecture.

Gibbs Smith, Publisher
P.O. Box 667, Layton, Utah 84041
www.gibbs-smith.com
alison@gibbs-smith.com
800-748-5439

Chelsea Green Publishing Company publishes books on sustainable living: straw bale, cob, rammed earth, cordwood, as well as books on treehouses, circle houses, and over a dozen books on renewable energy.

Chelsea Green Publishing Company
Post Office Box 428
Gates-Briggs Building #205
White River Junction, Vermont 05001
www.chelseagreen.com
info@chelseagreen.com
800-639-4099

Builders Booksource

Builders Booksource is an excellent source of building books.
1817 Fourth Street, Berkeley, CA 94710
www. buildersbooksource.com
service@buildersbooksite.com
Orders: 800.843.2028
Phone: 1-510-845.6874

Journal of Light Construction

JLC has a great website of no-nonsense books for contractors. Also, check out their $99 CD-ROM containing an indexed 17 years of the magazine.

186 Allen Brook Lane, Williston, VT 05495
www.jlconline.com
bookstore-cs@hanley-wood.com
802-879-3335

Searching Online

In addition to Amazon, try:
www.abe.com
www.fetchbook.info

More Great Books

Hey, space is limited! Here are a few more. (I often think of Ken Kesey's line in the '70s: "Take what you can and let the rest go by.")

The Cob Builders Handbook
Becky Bee
Groundworks, Murphy, OR

How to Build Low Cost Motorhomes
Louis C. McClure
Published by Ben Rosander

Sign Power
Kiyoko Semba and Kesaharu Imai
World Photo Press, Tokyo

Bernard Maybeck: Artisan, Architect, Artist
Kenneth H. Cardwell
Hennessy and Ingalls, Los Angeles

Cabins: A Guide to Building Your Own Nature Retreat
David and Jeanie Stiles
Firefly

Shelter's Worldwide Headquarters

OUR PRODUCTION STUDIO is built out of recycled wood from a nearby Navy base. It's on a half acre of land, along with our home, various outbuildings, and a vegetable garden. Due to digital technology, we not only can prepare a book for the printers right here, but we're hooked into the rest of the world. It's what writer Gene Youngblood forecast in the '60s, as "The Electronic Cottage."

Drawing by David Wills

ABOUT THE AUTHOR
What Led to Doing This Book

1947 12 years old, helped Dad build concrete block house in Sacramento Valley, Calif. My job: shoveling sand, gravel, cement into concrete mixer. One morning, got to nail down roof decking. This I liked!

1952–1954 Teen years, worked summers as carpenter, San Francisco docks; rough carpentry shoring cargo on outgoing ships. Learned what I could from other carpenters.

1947

1960

1960 After two years running USAF newspaper in Germany, returned to Mill Valley, California, went to work as insurance broker in San Francisco. In spare time converted old carport into a post/beam sod-roof studio. Did all cutting with hand saw. Shallow-rooted succulents planted on roof, white blossoms in spring. Liked building process, smell of wood, creating with own hands.

1963 Next project more ambitious. Used-wood, timber-frame house designed by architect friend — Japanese/Bernard Maybeck influenced. Post/beam frame, some 10'-high poured concrete walls. In over head, but got started and learned on job. Owner/builder perspective in learning to build. Have tried to maintain this outlook in reporting.

1963

1965 Hitchhiked across country, learned I had more in common with younger generation than my own, came back and quit insurance business, went to work as carpenter.

1966 Moved to Big Sur to work on job building large post/beam house (30' long, 8' × 22" used-redwood beams) on 400-acre ranch. Lived in chicken coop on ranch.

1966

1966

The '60s Magical cultural revolution that changed world going on. Mostly misunderstood these days. Artistic underground in San Francisco, early '60s. Beats fading artists of old world/hippies joyous, open, sharing/entirely different mindset. Wonderful few years (*before* "summer of love"). Non-conformity, dropping out, experimenting, searching, expanding awareness, looking for better ways to do things. Loving, exciting community on Haight Street, San Francisco, world headquarters for a few years.

All these things not so much new as being discovered for first time by millions of young Americans:

astronomy • astrology • meditation • Gurdjieff • Ouspensky • Zen Buddhism • Tarot • Kabbala • I Ching • dolphin consciousness • *Dune* & *Strangers in a Strange Land* • building your own house • organic gardening • ecological awareness • political activism • self-sufficiency • poetry • rock and roll • the blues • Native American culture • Ali Akbar Khan •

Beatles/Stones • Dylan • domes • LSD, marijuana, mescaline • Monterey Pop Festival • *Rolling Stone* • *Whole Earth Catalog* • *The Owner-Built Home* • *The Tassajara Bread Book* • viewing earth from space • Edmund Scientific catalog • L. L. Bean catalog • chickens by mail from Murray McMurray/and on and on . . .

1967

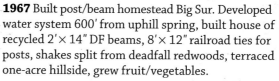

1967

1967 Built post/beam homestead Big Sur. Developed water system 600' from uphill spring, built house of recycled 2' × 14" DF beams, 8' × 12" railroad ties for posts, shakes split from deadfall redwoods, terraced one-acre hillside, grew fruit/vegetables.

1968 Buckminster Fuller influence, started building geodesic domes in Big Sur.

1968

1969 Got job coordinating building of 17 domes at hippie high school in Santa Cruz mountains. Experimented with geodesic domes of plywood, aluminum, sprayed foam, vinyl. Kids built own domes and lived in them. School became focus of media attention.

1969

1969

Left: First Whole Earth Catalog
Right: Domebook 2

1969–1970 Worked as Shelter editor for *Whole Earth Catalog.*

1970 First book published, *Domebook One.*
1971 Published *Domebook 2* — sold 175,000 copies and I was in publishing business.

1971 Bought half-acre lot small Northern California coastal town, built shake-covered geodesic dome — featured in *Life* magazine.

1971

1972 Decided domes didn't work, took *Domebook 2* out of print, disassembled and sold dome. Went in search of other (non-dome) ways to build — across U.S.A., Ireland, England — and *Shelter* (1973) was result.

1974

1974 Built stud-frame house using recycled lumber, doors, windows. Goal was to get shelter up quickly and have it be aesthetic and practical. Works great for us. Relief somehow to discover old ways can work best.

'80s, '90s Published series of fitness books, including *Stretching* by Bob Anderson.

34 years later: In 1994 visited sod roof house built in 1960 (see pic on p. 242) and met a happy trio living there: Jeff, Miranda, and baby Jesse.

2002–2004 Got back into the shelter (publishing) business. Operate out of recycled-lumber production studio in midst of vegetable garden, hooked into whole wide world via four Macintoshes.

I continue to travel and hunt for interesting shelter, maintaining layman's viewpoint; I love doing it!

Cameras

- Olympus OM-1s, full set of lenses
- Minox GT 35mm/*f*2.8 Leitz fixed lens
- Canon EOS A2E, 28–200 mm Tamron zoom
- Fujifilm 4700 digital 4.3 megapixel, incredible little camera
- Nikon 5700, 5.0 megapixel, 35–280 zoom

AND FINALLY...

Barn and vegetable garden in Connecticut

It's NOW EARLY OCTOBER 2003 and we're close to finishing *Home Work*. I have a big sign on my layout table saying "You don't have to include it all!" and I've been trying to adhere to that. Accordingly I have tons of material left over, the makings of another book similar to this one — sometime in the future.

With this book, we're going back to our roots: books on building. We're betting the farm on this one, and if it works, we have at least five years' worth of building books planned. (*See Coming Attractions, p. 245.*)

Please note: If you have any material to contribute to the next book in this series — photos, drawings, stories, advice, building adventures, insights, excitement — please get in touch with us.

Since I believe in filling up as much white space as possible, here are a few last-minute thoughts that didn't seem to fit elsewhere:

My PERSPECTIVE is that of an owner/builder, not architect or professional builder. In fact I don't trust experts. (Don't trust people with initials behind their names!) I've come around to looking for the simplest way to build as possible. For me, where I live, for speed, economy, and practicality, it's meant stud construction with mostly used wood.

Architecture
Pretty much all the new homes I see being built these days, especially in the wealthy Bay Area, are disasters. How can there be so much bad architecture? Don't get me started . . .

The Nomadic Carpenter
Lew, who is now our webmeister, scanner (and cover designer) was, a few years back, a carpenter. He carried all his tools in a Volkswagen van so he could do just about any job onsite. Small table saw, planer, chop saw, Skilsaw, Sawzall, jig saw, router, and several portable drills — along with other carpentry, plumbing, and electrical tools (and two dogs). He did soup-to-nuts — foundation work to cabinet work — and took everything to the job. It's a great idea for a young builder — remodeling, additions, fixing things up — learn to do it all yourself, keep it portable so you can go anywhere in the country. Work for people as a builder, not a contractor. Do the plumbing and wiring too. There's a great demand for solo builders like this, you'll always have work.

Outside/Inside
I've watched lots of couples building their dream homes. It's surprising the number of times in which the guy has an abstract idea what he wants to do, is thinking of what it will look like from the outside and the woman is thinking of what it will be like inside, how it will function for the life lived within. It's good when design decisions start with the latter, the design from within. Many couples break up when building a home, primarily because the job is more complex than it needs to be.

*A house without love
is not a home.*

—Hank Williams

Local Materials
In the early '70s, after I gave up on domes, my son Peter and I got on a charter flight to Ireland, crossed the Irish Sea, and started hitchhiking to visit friends who were living in a small town on the Thames near Reading. Coming across Wales, we got a long ride with a salesman; when he learned I was interested in building, he started pointing out buildings and showing us that each was built of materials from near the site. "You see the slate roofs, there's a slate quarry nearby . . ." and then, "Now the roofs are tile, because there's clay in the soil here . . ." As we travelled throughout England, it was striking: the thatched roofs in Norfolk, land of marshes and reeds; the sandstone walls in the Cotswalds, where the light tan colors blended perfectly with the surroundings; cob in Devon; flint in Sussex . . .

Don't Make It a Trip
I got involved in two building *trips* that lasted for about 12 years: heavy-timber post and beam frameworks, then rebounding to ultra-light geodesic domes, before discovering simple and conventional stud frame construction. If it were me, and I were building a house nowadays and I had to work for a living, I'd look for the simplest method around.

Home as Art? Be Sure That's What You Want to Do
If an artist makes a painting (or sculpture) that doesn't work out, it can be tossed. Art's a field where you can try things out. But a building is so big and complex and expensive that you're going to be stuck with it, maybe rationalizing unnecessary complexity. Yes, there's the other side of it: Lord knows a lot of people create delightful unconventional shelters. Look at my cousin Mike . . .

Choking Up
Burton Wilson told me this story: a young guy was working as a carpenter's apprentice. One day the carpenter noticed the kid was choking up on his hammer while framing, with his hand about about 2″ up from the end. "Hold it," said the carpenter. He took his pencil and put a mark on the hammer handle at the base of the kid's hand. Then he took his saw and cut the handle off at the mark. "You didn't need that part of the handle," he said.

So True
Sign at the San Marin Lumber Company in the '70s (and boy, did this resonate with me!):

If you didn't have time to do it right in the first place, how come you have the time to do it over?

*Either you is or you ain't;
Either you can or you cain't;
Either you will or you won't;
Either you do or you don't.*

—Rufus Thomas

I got to go to the top of the Golden Gate Bridge on a warm August evening in 2000. We went up inside the South Tower, then climbed a ladder to the top. What a thrill, growing up in San Francisco, then getting to the top of this beautiful bridge! This is the view looking towards Marin County.

CREDITS

Director of Production and Builder of this Book: Rick Gordon
Art Director: David Wills
Contributing Editor, Cover Design, Scanning: Robert Lewandowski
Photo Shoots: Janet Holden Ramos
Proofreading: Bob Grenier
Printing: C&C Offset Printing Co., Ltd., Shenzen, China;
    115 gsm multi-matte art paper,
    disk-to-plate process

Photo Processing over the years:
General Graphics Services, San Francisco
Chong Lee, San Francisco
Professional Color Labs, San Francisco
North Bay Photo Lab, San Rafael, California
Marin Filmworks, San Rafael, California

Rick, Lew, and David

This book was assembled by using:
Hardware:
Macintosh G4 and G3 computers, Nikon CoolScan 4000-ED film scanner,
Agfa Arcus II flatbed scanner, Epson 2200 Stylus Color Pro printer, Sony Artisan monitor, Gretag-Macbeth Eye One spectrophotometer

Software:
QuarkXPress, Adobe Photoshop, Microsoft Word, Nikon Scan,
Gretag-Macbeth ProfileMaker Pro, Compass Profile XT

Photo collages inspired by David Hockney
Layout inspiration from *Micro Architecture*, ed. Kesaharu Imai,
World Photo Press, Tokyo

With Help From:

Isidro Amora	Bob Easton	Lesley Kahn
Cafe Roma, San Francisco	Louie Frazier	Mike and Leda Kahn
Elise Cannon	Jack Fulton	Ian MacLeod
David Carriere	Molly Heriza, Kit Wong,	Mary Sangster
Karen Cross	Benny Chan, Esta Chan,	Bill and Athena Steen
Patrice Daley	& the folks at C&C Offset	Charlie Winton

Thanks to the three builders who got me started:
Lloyd Kahn, Sr. • Alec Fulton • Bob Whitely

PHOTO CREDITS

All photos by Lloyd Kahn, except those credited already, and:

2.	*Top:* Bob Gamlin
3.	*Top right:* Janet Holden Ramos
4.	*All photos except middle right:* from Louic Frazier
5.	*All photos except lower left:* from Louie Frazier
8–9.	*All black and white photos:* Janet Holden Ramos
10.	Leonard McLeod, Michael F. Bush
11–13.	Michael F. Bush, Barry Comber
14.	Leonard McLeod, Michael F. Bush, Nic Embleton
15.	*Lower left:* Ian MacLeod, others by "passers by"
21.	Mickey Castle and others
26.	John Silverio
28.	All by Bill Coperthwaite, except *top:* Dan Neumeyer
29.	All by Bill Coperthwaite, except *bottom left, and second down on right:* unknown photographers; *third down on right:* Daniel Taylor-Ide; *lower right:* Lloyd Kahn
38–39.	All from Karen Knoebber
42.	*Top left:* jean soum
44–51.	*All:* jean soum, except *top left*, page 51: Jean Michel Auriol
54–55.	Richard Perez
56.	Enrique Sancho Aznal
57.	Garry Crawford
59.	*Top left, lower left:* Janet Holden Ramos; *two middle left:* from Renée Doe
67.	*Lower right:* Burton Wilson

68.	*Both photos:* Burton Wilson
71.	Lester Walker
73.	Bill Steen
75.	*Top right:* Bill Steen
78–79.	Bill Steen
86–87.	Patrick Ironwood
88–89.	Kelly Hart
91–93.	Oscar Hidalgo
94–96.	Cookie and/or Rand Loftness
126.	*Middle, lower left:* Jack Fulton
127.	*Top right, third down on right:* Jack Fulton
128.	*Top:* Jack Fulton
130.	*Middle right:* from Maxine Page
134–139.	Steve Kornher
140–142.	David Greenberg
172.	*Center, lower left, lower right:* Jack Fulton
176.	*Rolling Homes photos:* Jane Lindz
178.	*Top two photos:* Rod Cathcart; *center left:* jean soum
179.	*Right, second down:* Ole Wik
180.	*Top right:* from Air Camping, Italy
181.	*Two Powerwagon photos:* H. L. Baggett
182–183.	Roger D. Beck
184.	From Nomadics Tipis
190–193.	D. Price
198.	From *Native American Architecture* and Nomadics Tipis
208.	*Top left, bottom right:* from Christoph Büchler
209.	All photos from Christoph Büchler
210.	John Welles
216–217.	Jim Macey
218.	From *Discovering Timber-Framed Buildings*

224–225.	Jim Macey
228.	*Middle left:* Joel Lundberg; *top right:* Richard Wachter
231.	*Middle left:* Godfrey Stephens; *bottom three photos:* Chuck Alexander
232.	Godfrey Stephens and friends

Drawings
Credits, unless credited on pages:

2–9.	Louie Frazier
12.	*Lower left:* David Wills; *lower right:* Ian MacLeod
13.	Ian MacLeod
25.	David Wills
26.	*Three drawings:* John Silverio
66–69.	All by Bob Easton
70–71.	All by Lester Walker
90–93.	All from *Bamboo — The Gift of the Gods*
99–101.	Eiko Komatsu
113.	Map by David Wills
128.	Map by Michael Kahn
186.	All from *Dwelling Portably*
207.	David Wills

NEW BOOKS FROM SHELTER PUBLICATIONS

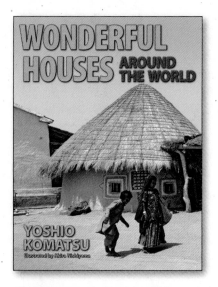

Wonderful Houses Around the World

By Yoshio Komatsu; illustrated by Akira Nishiyama

Paper: ISBN-10: 0-936070-34-X; ISBN-13: 978-0-936070-34-6 $8.95

Flexi: ISBN-10: 0-936070-35-8; ISBN-13: 978-0-936070-35-3 $14.95

7½" × 10" 48 pp.

For over twenty years, Yoshio Komatsu has travelled around the world, shooting photos of unique homes built out of natural materials. (See his work on pp. 98–105 of *Home Work*.)

   Wonderful Houses Around the World shows ten houses, accompanied by drawings of family life inside each house, and highlights what the children in each family do in their daily lives. This is a wonderful book for children in North America, showing them the very different homes and activities of their contemporaries in other parts of the world.

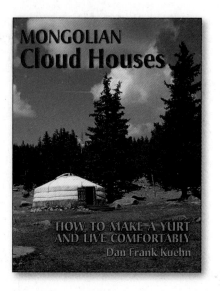

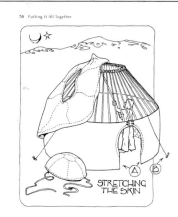

Mongolian Cloud Houses: How to Make a Yurt and Live Comfortably

By Dan Frank Kuehn

ISBN-10: 0-936070-39-0; ISBN-13: 978-0-936070-39-1; $16.95

7" × 9" 160 pp.

Friendly, easy-to-follow drawings take you step-by-step through the process of building a portable 13'-diameter yurt out of bamboo or willow, and canvas. Also included is the most up-to-date (2006) info on ready-made yurts, yurt building and covering materials, yurt web sites, and unique photos of present-day Mongolian herders and their nomadic dwellings. (See his work on pp. 188–189 of *Home Work*.)

   The appendix contains a wealth of information he's assembled on yurt designs, ready-to-erect yurts available from a wide variety of manufacturers, plus tools, material suppliers, yurt books, and videos. A wonderful last-minute addition is a section containing photos of modern-day *gers* in Mongolia.

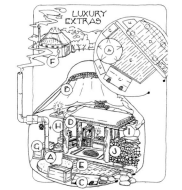